Douglas T. Gjerde
Christopher P. Hanna
David Hornby
DNA Chromatography

Douglas T. Gjerde, Christopher P. Hanna, David Hornby

DNA Chromatography

WILEY-VCH

Dr. Douglas T. Gjerde
12295 Woodside Drive
Saratoga, CA 95070
USA
Tel: 001-408 253 0927
gjerde@earthlink.net

Dr. Christopher P. Hanna
c/o HTS Biosystems
92 South Street
Hopkinton, MA 01748
USA
Tel: 001-508 435 4700
Channa8581@aol.com

Dr. David Hornby
Department of Molecular Biology
University of Sheffield
Western Bank
Sheffield S10 2TN
UK
Tel: 0044-1 142 224 236
Fax: 0044-1 142 762 687
d.hornby@sheffield.ac.uk

Cover
Lascaux cover design by Atelier Prisma, Tatjana Treiber, 76344 Leopoldshafen, Germany

Library of Congress Card No.: applied for
A catalogue record for this book is available from the British Library.

Die Deutsche Bibliothek –
CIP Cataloguing-in-Publication-Data
A catalogue record for this publication is available from Die Deutsche Bibliothek.

Printed in the Federal Republic of Germany.

Printed on acid-free paper.

Typesetting Hagedorn Kommunikation, Viernheim
Printing betz-druck gmbh, Darmstadt
Bookbinding J. Schäffer GmbH & Co. KG

ISBN 3-527-30244-1

Contents

Preface *XI*

Acknowledgement *XIII*

About the cover *XV*

1 **Introduction** *1*
1.1 General Background *1*
1.2 Short Historical Review of the Chromatography of Nucleic Acids *6*
1.3 Terms and Definitions *8*
1.4 Scope and Organization of This Book *10*
References *11*

2 **Instrumentation and Operation** *14*
2.1 Introduction *14*
2.2 General Description of the DNA Chromatograph *15*
2.3 Detailed Description of the DNA Chromatograph *16*
2.3.1 The General Instrument and Materials *16*
2.3.2 Dead Volume *17*
2.3.3 Degassing the Eluent *18*
2.3.4 Pumps *19*
2.3.5 Gradient Formation *21*
2.3.6 Pressure *23*
2.3.7 Autosampler Injector *23*
2.3.8 Separation Column *25*
2.3.9 Column Protection *26*
2.3.10 Column Oven *27*
2.3.11 Detection *28*
2.3.11.1 Selective vs. General Detection *28*
2.3.11.2 Ultraviolet-Visible Detectors *29*
2.3.11.3 Fluorescence Detector *31*
2.3.11.4 Mass Spectrometry Detection *34*

2.3.12 Data Analysis 37
2.3.12.1 Size Analysis 38
2.3.12.2 Peak Shape or Pattern of Peaks 38
2.3.12.3 Quantification 39
2.3.13 Fragment Collection 40
References 41

3 Chromatographic Principles for DNA Separation *42*
3.1 Introduction 42
3.2 Comparison of Chromatography and Gel Electrophoresis 45
3.3 Basic Chromatographic Considerations 48
3.3.1 Retention 49
3.3.2 Retention Factors 50
3.3.3 Peak Width 50
3.3.4 Plate Theory of Chromatography 51
3.3.5 The Rate Theory of Chromatography 55
3.3.5.1 General Considerations 55
3.3.5.2 Extra Column Effects 57
3.4 Reverse Phase Column Packing Materials 57
3.4.1 Types of Materials 57
3.4.2 Polymeric Resins 59
3.4.2.1 Substrate and Crosslinking 59
3.4.2.2 Porous and Nonporous Resins 59
3.4.2.3 Monolith Polymeric Columns 62
3.4.2.4 Functionalization of the Polymer 62
3.4.3 Silica-based Materials 64
3.4.3.1 General Description 64
3.4.3.2 Functionalization 65
3.5 Reverse Phase Ion Pairing Chromatography 66
3.5.1 Principles 66
3.5.2 Temperature Modes of DNA Chromatography 71
3.5.3 Effect of Metal Contamination 72
3.6 Ion Exchange Materials and Separation Mechanism 75
3.6.1 Polymer-based Anion Exchangers (Anex) 75
3.6.2 Silica-based Anion Exchangers 77
3.6.3 Basis for Separation 78
References 79

4 DHPLC *81*
4.1 Introduction 81
4.2 Practice of the Technique 83
4.2.1 Melting Phenomena and Domains 83
4.2.2 Temperature Prediction 85
4.2.3 Primer Optimization and Clamping 88
4.2.4 PCR Fidelity 91

4.2.5 High-sensitivity DHPLC Determinations *92*
4.2.5.1 Chromatographic Resolution between Heteroduplices and Homoduplices *93*
4.2.5.2 Mass Sensitivity for the Resolved Heteroduplices *93*
4.2.5.3 PCR-induced Background *95*
4.2.5.4 Example Application: Detection of Varying Levels of k-ras Alleles *96*
4.3 Review of DHPLC Publications *98*
4.4 Conclusions *103*
References *105*

5 Size Based Separations *108*
5.1 Introduction *108*
5.2 Fundamental Developments *108*
5.3 Calibration *110*
5.4 Applications *114*
5.4.1 Primer extension and near-size-based separations *114*
5.4.2 LOH and other size-based genotyping techniques *115*
5.4.3 Size Based Purification Procedures *115*
References *116*

6 Purification of Nucleic Acids *118*
6.1 Introduction *118*
6.2 System Dead Volume *119*
6.3 Cleaning *120*
6.4 Testing the Instrument Operation *122*
6.5 Calibration and Separation Conditions *122*
6.5.1 Internal and External Calibration *122*
6.5.2 Isocratic Elution *123*
6.6 Software Collection Methods *132*
6.7 Recovery of Material *133*
References *134*

7 RNA Chromatography *135*
7.1 Introduction *135*
7.2 Biological Extraction of RNA *137*
7.3 Size Based RNA Separation *139*
7.4 Separation of Cellular RNA Species *141*
7.4.1 Separation of Messenger RNA from Ribosomal RNA *141*
7.4.2 Analysis of Transfer RNA *143*
7.5 Chromatography and Analysis of Synthetic Oligoribonucleotides *145*
7.6 Application of RNA and DNA Chromatography in cDNA Library Synthesis *149*
7.7 Analysis of Gene Expression by RNA and DNA Chromatography *152*
7.7.1 DNA Chromatography Analyses of RT-PCR and Competitive RT-PCR Products *152*

7.7.2 Alternative Splicing *154*
7.7.3 Differential Messenger RNA Display via DNA Chromatography *155*
References *158*

8 Special Techniques *160*
8.1 Introduction *160*
8.2 Analytical and Preparative Enzymatic Cleavage of DNA *161*
8.3 Analysis of DNA Methylation *164*
8.4 Nucleic Acid Enzymology *171*
8.4.1 Telomerase Assays *171*
8.4.2 Polynucleotide Kinase Assays *173*
8.4.3 Uracil DNA Glycosylase Assays *174*
8.5 Protein Nucleic Acid Interaction Mapping: "Footprinting" *175*
Method Section *177*
8.6 Nucleic Acid Tagging *180*
8.7 DNA Chromatography with Intercalating Dyes *181*
References *182*

9 Looking Forward *184*

Appendix 1 Glossary of Terms *187*

Appendix 2 System Cleaning and Passivation Treatment *207*
A2.1 Background Information *207*
A2.2 Reagents *211*
A2.3 Preparation of the System *211*
A2.4 Passivation of System *211*
A2.5 Equilibration of System *212*
A2.6 Passivation of Injection Port and Injection Needle *212*
References *213*

Appendix 3 Frequently Asked DHPLC Questions *214*
A3.1 What are the various methods for DHPLC temperature selection? *214*
A3.2 Does having the optimum oven temperature mean that I will get the optimum resolution of the heteroduplex and homoduplex species? *215*
A3.3 I do not have the complete sequence of my fragment but I want to scan for mutations. Is this still possible by DHPLC? *216*
A3.4 Will DHPLC detect both heterozygous and homozygous mutations? *217*
A3.5 I have a sample population to scan for mutations and need to be certain that I find all mutations present. What are the factors affecting the accuracy of DHPLC and how should I approach the problem? *217*
A3.6 What are the minimum and maximum fragment sizes I can analyze by DHPLC? *218*

A3.7 I need to screen a large number of samples. What is the quickest way to do this? 219
A3.8 I want to create a general SNP map but do not need to find every mutation. What is the best strategy? 219
A3.9 What will happen if I have more than one mutation in a fragment? Does each mutation or combination of mutations give a unique chromatographic pattern? 220
A3.10 What about the converse? Does a particular chromatographic pattern indicate a particular mutation? 220
A3.11 I have a biological system where it is likely that a mutation if present can have a heterduplex concentration of less than 50 %. What is my best approach to this problem? 221
A3.12 Is it possible to use DHPLC in a diagnostic setting? 222
References 223

Index 225

Preface

The term DNA Chromatography is defined as a high performance, automatic separation and purification of DNA by high performance liquid chromatography (HPLC). Until recently, HPLC separation of DNA was too slow or DNA fragments were too poorly resolved to be of much use to the molecular biologist. Furthermore, most of the HPLC journal publications on DNA separation published by analytical chemists were written about technology that was not relevant to the needs of the researcher. For example, a typical publication might demonstrate the column and conditions to separate poly-T oligonucleotides up to 20 nucleotides in length. Of course, demonstrating the separation of a mixture of poly-T oligonucleotide is of little interest to the molecular biologist because there is not much biological necessity to study this type of DNA. The analytical chemist simply did not understand the problems facing the molecular biologist. The analytical chemist knew that DNA separations were important, but did not understand how the molecular biologist needed to perform the separation or why and how the information would be used. This is not to disparage analytical chemists. It is only to say that there exists a chasm of understanding between analytical chemists and molecular biologists.

This chasm must be crossed mainly because the needs of molecular biology science are changing rapidly. The tools that are needed to understand molecular biology are analytical tools. In order to understand and ultimately control the molecular basis of life, analytical experiments must be designed and implemented, analytical tools must be used, and analytical information must be generated and studied. Because of its inherent analytical nature, DNA chromatography has the potential in becoming one of the leading analytical tools used by the molecular biologist.

Who is to cross this chasm between analytical science and molecular biology? Some may say it must be the analytical chemist. They must explain the need for certain standards and methods for performing separations. But the onus cannot be completely on the chemist. The molecular biologist must recognize that there are new needs and standards that must be applied to their work. Thus, while the analytical chemist must teach their art to the molecular biologist by using terms that are clearly defined, the molecular biologist must teach the analytical chemist of their needs. Their needs are expressed in the objectives or goals of their research, how their experiments are designed and how the results might be used. It is the intent of the authors that by introducing DNA Chromatography, this book will provide some of the means to cross this chasm.

Douglas T. Gjerde
San Jose, California

Chris Hanna
Cambridge, Massachusetts

David Hornby
Sheffield, UK

Acknowledgements

The authors would like to acknowledge and thank many people who contributed or helped with the writing of this book. Leon Yengoyan of San Jose State University read major portions of this book and made many useful comments. The authors hope that they have used the comments and suggestions wisely. The authors are grateful to Joann Walters who made several helpful comments on Chapter 4 DHPLC and who also is the author of Appendix 3: Frequently Asked DHPLC Questions. The authors thank Maryam Matin and Mark Dickman who developed many of the RNA and DNA applications and helped assemble much of the information that is reported in this book. They are both extremely talented and the authors appreciate their dedication. The authors thank and acknowledge the contribution of Robert Haefele to Appendix 2, the section on system cleaning and passivation.

Jeffrey Sklar of Harvard Medical School has provided many valuable insights regarding DNA Chromatography technology as a whole, and all of the authors are grateful for his efforts through the years. The authors thank Günther Bonn of Leopold-Franzens University Innsbruck for his support of the technology and for the many hours of discussions. Bert Volgelstein of Johns Hopkins Medical School provided the k-ras samples described in Chapter 4 and made very useful comments to the analyses that were performed.

The authors are grateful to Arezou Azarani for the information provided on the denaturation of RNA and to John Brady and Bill Walker who were very helpful in providing many of the patent references and figures. Pam Smith and Tiffany Nguyen are thanked for their contributions to the organization of the literature and their help with the logistics of the project. The authors would like to thank their friends and colleagues for the many hours of enjoyable discussions.

Finally, the authors would like to acknowledge and thank their families. They are the most important parts of our lives. Their support and encouragement did not go unnoticed and the authors express their love to them.

About the Cover

The cave paintings found in France, Northern Spain and other places represent some of the earliest forms of enduring art produced by our species. They tell us about ourselves but they also prompt us to ask questions.

Who were the people that made these paintings? At the dawn of civilization – over 35,000 years ago – it is possible Cro-Magnon and Neanderthal man lived at the same time? Did they interact to the extent that we are descendents of both species?

Why are many of the animals depicted in the paintings now extinct? Is extinction something that man can avoid?

One of the ways researchers are attempting to answer these questions is through DNA analysis. DNA is enduring. We are the same as the artists who painted these drawings. Our DNA will tell us our past, who we are and perhaps point the way to our future.

1 Introduction

1.1 General Background

The knowledge base of molecular biology has been built upon the use of analytical tools. Sequencing, allele specific amplification, quantitative PCR, ligase chain reaction, etc. are all analytical methods that, if used properly, allow molecular biologists to design experiments and interpret information and contribute to the base of knowledge. The success of any particular method depends on how well it can perform its analytical function. How reproducible is the method? Does the method have interferences that could result in either false negative results or false positive results? Is the test robust? Will the analytical test function in the same manner every time it is used by an analyst or by several analysts?

There has been rapid growth over the last ten years in the number of analytical tools available for the analysis of nucleic acids. The large majority of these tools are based or depend upon gel electrophoresis. More specifically, analytical methods have been developed around the assumption that gel electrophoresis is required for the development of the information being sought. Separation science is one of the mainstay tools used in chemistry and biology. In particular, technological advances that have facilitated the analysis and preparation of pure biopolymers have been central to the development of modern molecular biology. Gel electrophoresis has allowed the separation and purification of a wide variety of nucleic acid molecules.

Slab gel electrophoresis is the cornerstone tool of the molecular biologist. It has been in use for almost 60 years, and is on the lab bench of virtually all researchers. The tool will separate DNA fragments and with visual scanning options, the researcher can determine the size and amount of DNA separated. High resolution purification of a particular band can be accomplished by cutting out with a scalpel or razor blade, then the particular band of interest is extracted from the gel matrix.

As the speed of throughput in contemporary experimentation becomes ever more demanding, the need to automate has been acknowledged to be of prime importance. Due to many of the tedious aspects related to technique, there have been attempts to automate the slab gel electrophoresis process. There are very few options for simple sample introduction, separation, detection and automatic collection of separated materials. Pre-cast gels with automated visual scanning detection devices have helped the technique.

The introduction of capillary gel electrophoresis has been successful in circumventing many of the drawbacks of slab gel electrophoresis. Materials separated by capillary gel electrophoresis are not easily purified and collected from the capillary instrument and therefore the instrument falls somewhat short as a suitable alternative for the molecular biologist.

There will always be a strong need to analyze nucleic acids for purity and size and to prepare pure materials. In an effort to avoid the complex tasks of gel automation, a variant of HPLC known as DNA Chromatography has been developed. HPLC is a highly automated technology. It employs automatic sample introduction, separation, and detection and even automated collection of separated samples. Depending on the column and separation conditions, large quantities of nucleic acid fragments can be separated, purified and collected. The instrument is computer controlled.

Simply stated, DNA Chromatography is the high performance separation of nucleic acids by high performance liquid chromatography (HPLC). High performance implies that modern standards of instrumentation are met, and separations of high purity are achieved. In the case of DNA, the HPLC instrumentation is computer controlled. The separation conditions are computed and implemented through input of the DNA sequence and the desired type of separation. DNA may be collected automatically for further downstream processing such as cloning, PCR, sequencing, etc. The separations are performed usually in less than 10 minutes and, in many cases, single base pair resolution of the DNA is achieved.

An example of an application of DNA Chromatography is illustrated in Figure 1.1. Competitive quantitative, reverse transcription PCR (Q-RT-PCR) is a sensitive method for measuring trace mRNA. A quantity of sample is mixed with a known amount of competitor to act as an internal standard. The competitor has almost an identical sequence to the unknown except for an (approximately 20 base) insertion somewhere in the fragment. The two species are reverse transcribed to cDNA and the sample is amplified by PCR (at equal efficiency for sample mRNA and competitor). The concentration of the sample mRNA is calculated using the concentration of the competitor standard and the fragment measurements. Previously, the measurement of the fragments had been done by gel electrophoresis analysis. Peter Doris and his coworkers worked out the method to perform the analysis by DNA Chromatography [1, 2]. They found that this allowed a much more accurate measurement than previous methods mainly because the heteroduplex complexes of the sample and competitor could now be reliably resolved. Gel methods usually don't have the resolution to separate the heteroduplex species from the sample and competitor fragments leading to errors in the calculation of the mRNA concentration. The DNA Chromatography method can be applied to high-throughput research and clinical facilities, especially where high sensitivity (for low expressed forms of mRNA) and high accuracy of mRNA measurements are needed. (Details of the method are described in Chapter 7).

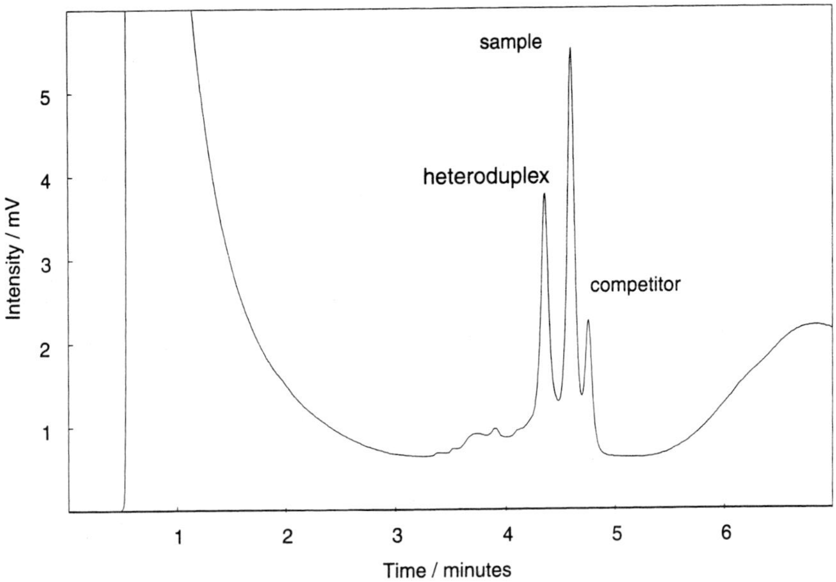

Figure 1.1. Separation of Q-RT-PCR target 204 bp sample fragment, 228 bp competitor DNA fragment, and heteroduplex on a DNASep® column and WAVE® System: Eluent A: 0.1 M triethylammonium acetate, pH 7.0 (TEAA); B: 0.1 M TEAA, 25 % acetonitrile (from Ref. [2] with permission).

There have been many different approaches published on the analysis and purification of nucleic acids by chromatography. A short historical review is described in the next section. The fundamental technology leading to modern DNA Chromatography was first described by Günther Bonn, Christian Huber and Peter Oefner in 1993 [3–5]. They showed rapid, high resolution separations of both double-stranded and single-stranded DNA. The separations were performed usually in less than 10 minutes and, in many cases, single base pair resolution was achieved. This form of HPLC analysis is largely (though not entirely) based upon the unique separation properties of a non-porous polystyrene–divinylbenzene polymer bead that has been functionalized with C-18 alkyl groups. An alkylammonium salt is added to the eluent and forms neutral ion pairs when a DNA sample is introduced into the HPLC instrument. A gradient of acetonitrile solvent separates the DNA fragment with the smaller fragments coming off the column first and then larger fragments eluting off the column and traveling through the detector. Figure 1.2 shows a separation of double-stranded DNA that can be achieved.

The separation occurs in such a manner that classic gel-based separation are mimicked. Bonn, Huber and Oefner showed that the DNA separations were performed according to the size of the fragment just are they are in gel electrophoresis. For double-stranded DNA, the sequence does not contribute to the retention of the fragment. Figure 1.3 taken from their work demonstrates this with a plot of retention time vs. fragment size for a number of different fragments. Various plas-

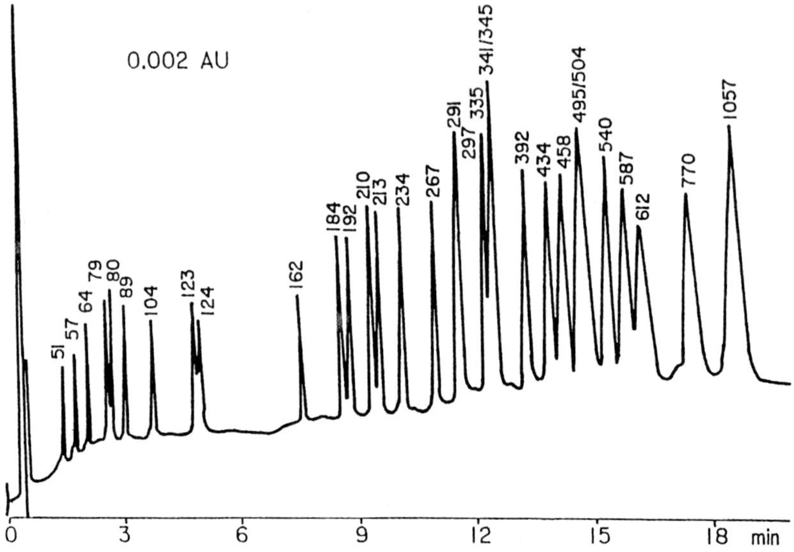

Figure 1.2. High performance DNA Chromatography separation of double-stranded DNA using a DNASep® column 50 x 4.6 mm. Sample was a mixture of BR322 *Hae* III restriction digest and _174 Hinc II restriction digest. Eluent and Gradient: A: 0.1 M TEAA; B: 0.1 M TEAA, 25 % acetonitrile, 35 to 45 % B in 2 min., 45 % to 57 % in 10 min., 57 % to 61 % in 4 min., 1.0 mL/min flow rate, UV detection at 254 nm (from Ref. [4] with permission).

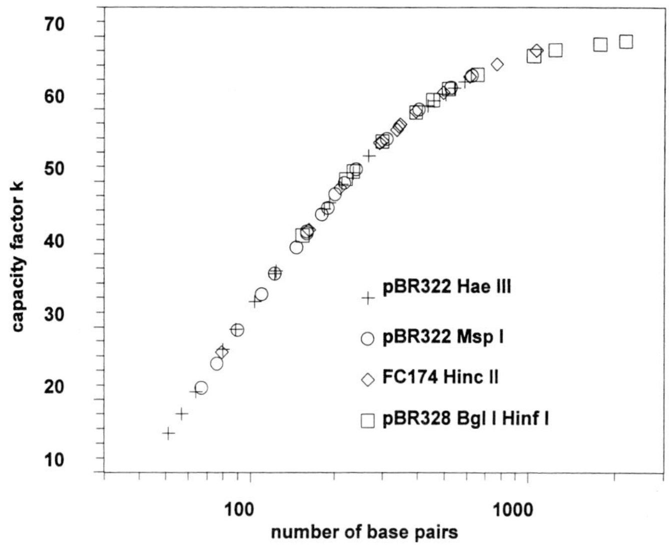

Figure 1.3. The retention times of various plasmid digest fragments are plotted according to size. The fragments have different sequences, but are separated according to their size using a DNASep® column 50 x 4.6 mm. Eluent and Gradient: A: 0.1 M TEAA; B: 0.1 M TEAA, 25 % acetonitrile, 0 % to 100 % B in 30 min., 1.0 mL/min flow rate, UV detection at 254 nm (from Ref. [41] with permission).

mids were digested with a number of different enzymes so that a mixture of DNA fragments were generated. Since the sequence of the plasmids are known, the effect of DNA sequence can be shown. In the plot, all of the retention times of the various fragments fall on a line drawn through the data. This shows that the retention of DNA is dependent on fragment size, and is independent of fragment sequence.

One major difference between chromatography and electrophoresis is the use of gradient elution in chromatography. The content of the fluid (eluent) that is pumped through the column can be changed in real time as the separation is proceeding. This is called a gradient process whereby the "strength" or "eluting power" of the eluent is increased as the separation is developed. Small fragments are eluted first and then the organic solvent is increased and larger and larger fragments are eluted. The time required for the entire separation depends on the "eluting program" that is employed. A rapid increase in the stronger eluent will elute the fragments faster. But if a rapid program is used, a separation with lower resolution of peaks could result. Conversely, a slow gradient process will produce the highest resolving conditions and greatest separation of peaks. However, this carries the penalty of a longer separation time. If the DNA Chromatographic system is well designed and a high efficiency column is used, then rapid gradients will produce suitable and useful separations.

Conversely, electrophoresis must rely on the conditions selected for the entire separation. The resolving conditions are not changed once the run is initiated; thus, whatever conditions were first selected must be used for the entire process. The ability to separate with a fluid gradient gives DNA Chromatography a tremendous advantage over gel electrophoresis.

Another advantage of DNA Chromatography is the ability to detect the DNA fragment without fluorescence detection. Tags need not be used and the fragments may be detected directly at the sub nanogram level by UV automatic detection. Of course fluorescence detection could also be used provided fluorescence tags are added to the DNA. Use of this detection method decreases the amount of DNA that can be detected by a factor of 10 to 100 (or even 1000 in some reported cases).

The use of column temperature to control separations in DNA chromatography was described in 1996 through the insights of Oefner and Underhill [6–9]. They demonstrated that DNA Chromatography possessed unique properties enabling the separation of DNA based on its relative degree of helicity. Heteroduplex DNA has a lower melting point than homoduplex DNA. The retention of single-stranded DNA is lower on the column than double-stranded DNA, thus heteroduplex DNA melts more easily, consequently it comes off or elutes from the column earlier in the separation. This technique is called denaturing HPLC (DHPLC). In short, they developed a method of DNA chromatography that is analogous to denaturing gradient gel electrophoresis (DGGE).

Gel electrophoresis is a powerful tool. But, depending on the problem, DNA Chromatography has many advantages. The development of high resolution chromatographic methods to replace many of the tasks now performed by gel electrophoresis almost seems inevitable.

1.2 Short Historical Review of the Chromatography of Nucleic Acids

Much of the early liquid chromatography work published on the separation and purification of nucleic acids may seem crude by comparison to modern capabilities. But it would be a mistake to completely discount this early work. Many of the research reports describe sample preparation schemes, experimental design, separation schemes, and down stream processing of materials that are still important. Certainly, the advanced technologies in use today are built upon the careful work that preceded these advances. Still other early work is remarkable in its foresight and quality. The short review in this section is not meant to be complete and only some of the important references are cited. The section is intended to give a brief overview of the types of work that have been performed.

The development of the high performance separation of nucleic acids has had a long history beginning with the early days of liquid chromatography. The progress has proceeded at a slow pace even as high quality separations of other biomolecules such as proteins were being developed. DNA Chromatography has unique requirements for clean instrumentation and well designed columns and cleaning methods as described in this book. But single-stranded DNA has less stringent requirements than double-stranded DNA, and the best publications were based on work performed with single-stranded DNA.

J. Thompson and coworkers published many excellent papers both in the form of review articles [10–15] and research papers [16–18]. The review articles published in 1986 and 1987 are a series of 6 publications, each dealing with some aspect of nucleic acid separation. In the first review (I) paper, Thompson presented a brief overview of the development of chromatographic stationary phases for liquid chromatography. He stated that parallel improvements in both silica and polymer-based resins were helping to change liquid chromatography nucleic acid separations from classical techniques to more advanced HPLC techniques. He discussed many different types of chromatographic separations, (some of which will be discussed in this section), ion exchange, and ion-pairing, reverse phase separations (the basis for much of this book). Reviews II, III, and IV discussed the isolation, purification, and analysis of single-stranded DNA, plasmids and double-stranded DNA. The separations are quite good, but still mostly slower, and lower resolution than what can be achieved today.

Review V describes the use of affinity chromatography. Double-stranded nucleic acid is held together by hydrogen bonding of the two pyrimidine bases; thymine (T), and cytosine (C) and the two purine bases; adenine (A) and guanine (G). The basis for forming the double-stranded DNA is that sequence A will always hydrogen bond with T, and G will always hydrogen bond with C. In affinity chromatography, a short single-stranded of DNA is attached to a solid support (perhaps 30 bases). The sample containing the target nucleic acid fragment is passed through a column containing the support. Only the sample nucleic acid that finds an exact match with the complementary fragment will become attached to the solid support through hydrogen bonding. Other sample matrix nucleic acids and sample matrix materials are washed through the column. And then using solvent or heat, the target nucleic acid fragment is denatured from the support and collected. An example of this technology is a support containing a bound oligo (dT) polymer used to purify mRNA through its poly A tail. Much of the review discusses the various supports to which the binder nucleic acid is attached including cellulose, nitrocellulose, methylated cellulose, cellulose acetate, and of course silica and polymers. The review also discussed the conditions for hybridization and denaturation of the target material.

There are many proteins that are attracted to specific DNA sequences. In Review VI, Thompson et al discussed using the same sort of substrates to study DNA-binding proteins. Many proteins that function in association with cellular nucleic acid recognize these polymer as a substrate and bind tightly under various conditions. Binding under different variables such as different ligand recognition sequences, ionic strength, pH, divalent cations and non-specific competitior DNA can be studied to determine the strength of the binding.

There have been many studies of different ways to separate DNA and RNA. A few of these studies are listed here. Size exclusion of double-stranded DNA restriction fragments was shown to be an accurate method for size determination of unknown DNA. The separation is based on a hydrated gel packing particles. Smaller fragments can penetrate the column packing and travel through the column more slowly. Larger fragments are excluded from the packing and travel through the column faster. The separation is the inverse to gel electrophoresis where the basis for separation is the restriction to travel of the larger fragments through the medium. Ellegren and Laas used Superose 6 and Sephacryl® S-500 column packings and a 0.12–0.2 M solution of sodium chloride eluent to produce some nice size-based separations [19]. The major drawback to the technology is that the separation times can be quite long.

Some of the first work to separate larger double-stranded DNA was performed using inorganic hydroxyapatite, a form of calcium phosphate [20, 21]. Particles of this material are crushed sized and packed into columns. At a low eluent pH, these materials will act as anion exchangers. Hydroxyapatite column were used primarily for purification. For example DNA is purified from RNA or single-stranded DNA is purified from double-stranded DNA. However, the phosphate buffers used for elution are more difficult to remove from the nucleic acids.

The synthesis and development of column materials for HPLC is difficult work. In an effort to develop faster, more versatile chromatographic materials, several research groups developed coated anion exchangers. These ion exchangers are prepared by coating a neutral column material with an insoluble organic material that contains an ion exchange group. This early work was called reverse phase chromatography, but was actually ion exchange chromatography. The coating molecule contains a nonpolar part that is attracted to the nonpolar packing material and also contains an ionic part that will undergo ion exchange chromatography with the nucleic acid. Although useful for transfer RNA [22–24] and plasmids and other DNA [25], the early column materials were somewhat difficult to use due to bleeding of the coating material from the column. Later a more stable material called the RPC-5 analog was developed and several applications were reported [26, 27].

Coated materials are versatile, but always have the disadvantage of being unstable. Ion exchangers, where the groups are chemically bonded to the substrate, have proved to be much more rugged and reliable [28, 29]. The major problem with anion exchange has been that the separations are not size based. Commonly, separation reversals will occur, e. g. larger fragments will sometimes elute before smaller fragments in a separation [30, 31]. Nevertheless, these publications show that ion exchange can be extremely rapid and useful when using modern non-porous anion exchangers – especially for the analysis of PCR products [32–35]. The resolution of double-stranded DNA is good, the separations can be performed in a few minutes and even very large fragments of several thousand base pairs can be separated.

Ion pairing, reverse phase chromatography is the basis for most of the applications described in this book. Early work showed that the separations of double-stranded and to a limited extent, single-stranded were size-based and not sequence based [36–39, 13]. With the introduction of high performance columns [3–5, 40–43], excellent DNA Chromatography separations have been achieved. This technology has been applied to a wide variety of applications including quantitative PCR [44–46], cloning purification [47], RNA analysis and purification [48–49], viral diagnostics [50], short tandem repeat analysis [51], population analysis [52], primer extension [52–55], pooling to determine allele frequency [56], and many others. Many of the references to date have involved the discovery and analysis of mutations by DHPLC (Chapter 4).

1.3 Terms and Definitions

Many of the literature references use the term DHPLC to cover all types of nucleic acid separations. In some of these papers, DNA is partially denatured. In other papers the DNA may be fully denatured or not denatured at all. Careful use of terminology is important. In this book, DHPLC refers to only partial denaturing heteroduplex analysis.

Other literature references will use the term ion pairing, reverse phase HPLC. This terminology is confusing and non specific because there are many methods that use ion pairing, reverse phase HPLC have nothing to do with DNA. As we have noted, this book uses the term DNA Chromatography. This can be performed in non denaturing mode or the full or partial denaturing mode if further clarification is needed. DHPLC is a part of the larger technique of DNA Chromatography. The terms ion pairing and reverse phase are used in this book as descriptive terms where needed.

The separation of RNA also falls into the broad category of DNA chromatography. There are strong similarities on how DNA and RNA are separated and collected. But there are differences as well, and no doubt nomenclature in this area will evolve. The term oligonucleotides, generally refers to single-stranded DNA. Oligoribonucleotides is used as the term for (single-stranded) RNA, although oligodeoxyribonucleotides is becoming accepted to distinquish single-stranded DNA from RNA oligonucleotides.

Throughout this book, many other technical terms are used. Since this is an analytical science book, many of the terms are analytical in nature. Yet, DNA Chromatography is meant to be practiced in the molecular biology world. Thus, it is important that the analytical language continues to adapt to the molecular biology world. Of primary consideration has been to use the term more likely to be used by the molecular biologist rather than the analytical chemist. For example, an analytical chemist would use the term standard, rather than control, to describe the material used to calibrate an instrument or a particular instrument. Both terms are used in this book.

To the analytical chemist, the term sensitivity is simply a ratio of the analytical detection signal to the analyte amount (or sometimes the analyte concentration). To the molecular biologist, the term sensitivity might refer to the ability of a particular technique to detect the presence of a particular mutation in the presence of a high wild type background. In cases where a term can be misunderstood or misused, the molecular biology use will be used if possible and definitions will be included in the text. For the analytical chemist, the concept of analytical instrument noise, noise contributed from the method itself, and the ability to measure a signal above this noise level is the subject of countless analytical chemistry papers.

The answer being sought in molecular biology can often be framed as either "yes" or "no". To an analytical chemist, it is not enough to say just "yes" or "no". The analytical chemist wants to know how much of something must be present before the answer be "yes". And if the answer is "yes", how much exactly is present? This may seem unnecessary to the molecular biologist but once this kind of information is understood and used, the ability to understand the results of the analytical tests become much more useful.

The terms, precision and accuracy, are easily misused or interchanged by the non-analytical chemist, and yet they have quite different meanings. Precision is the ability to repeat a result. Accuracy is the ability to determine the correct answer. It is quite possible to obtain the same answer many times, but the answer may still be incorrect.

A glossary section has been included in Appendix 1 in the back of the book. At least half of the battle of learning a new discipline is learning and understanding the language of that discipline. The molecular biologist reader is encouraged to read through the glossary section several times until the terms are understood. The analytical chemist who reads this book should also consult the glossary to make certain that their use of the jargon is correct for molecular biology and DNA Chromatography.

1.4 Scope and Organization of This Book

This book is designed to meet the needs of the community of biologists for whom HPLC has been something to be avoided. Here we set out to explain the fundamental principles of chromatography as related to the separation and analysis of nucleic acids. In addition, we wish to provide a comprehensive reference source of information for the practicing DNA chromatographers.

Instrumentation and operation of DNA chromatography are described in Chapter 2. Although the exact form of the instrumentation used can differ, there are factors that are constant from instrument to instrument. Enough detail is described so that some limited amount of trouble shooting can be performed. Also, maintenance of the instrumentation is described. Additional information for procedures regarding system cleaning and passivation are listed in Appendix 2. Chapter 3 is dedicated to the description of the separation column and the chromatographic process. The different parameters that affect DNA Chromatography and their control are described.

Chapter 4 describes a major application of DNA Chromatography: mutation detection by denaturing HPLC (DHPLC). The parameters that affect the practice of these methods as well as their optimization are described. Several common questions and answers with regard to the practice of DHPLC are listed in Appendix 3.

Several size based applications of DNA Chromatography are described in Chapter 5. Just as gel electrophoresis is used on a wide variety of applications, DNA Chromatography can also apply. The strengths of the quantification power of the technique are described.

Collection and purification of nucleic acids are described in Chapter 6. The purified materials may be used for downstream processing such as sequencing or PCR.

The applications of the technology to RNA are described in Chapter 7. RNA is extremely unstable and difficult to analyze. The methods needed to separate this material are described.

Chapter 8 describes several of the special techniques that can be employed to tag DNA or measure the interaction of DNA with other materials. Finally, looking toward future developments is described in Chapter 9.

It is clear that DNA Chromatography offers substantial opportunities for biologists and chemists who wish to explore more fundamental and applied aspects of the genetic material. Mutation detection (heteroduplex detection) provides one of the major focal points of the book illustrating the potential of DHPLC for high throughput and high sensitivity genetic analysis. Other applications are discussed with the aim of stimulating the development of novel applications of the technology in the life sciences.

This book aims to provide a comprehensive resource for nucleic acid chromatographers, to educate the novice practitioners, and whet the appetite of the non-believers. With the introduction of resins capable of delivering the resolving power of polyacrylamide electrophoresis gels, coupled with automated sample delivery, detection, analysis, and collection, DNA Chromatography has come of age. Since the pioneering experiments of Bonn and coworkers, DNA Chromatography has become commercially available, and has given rise to the publication of more than 100 refereed publications. It is the goal of the book to compile this published information, and also to interpret a fair amount of it based on the current state of knowledge. This monograph also attempts to place DNA Chromatography in the larger context of other nucleic acid analysis approaches. Lastly, the monograph attempts to give a view of the future for DNA Chromatography and areas for its future development.

References

1. A. Hayward-Lester, P. J. Oefner, S. Sabatini, and P. Doris, Accurate and absoulte quantitative measurement of gene expression by single-tube RT-PCR and HPLC, Genome Res., 5, 494, 1995.
2. K. Munson, R. Haefele, and P. Taylor, Competitive quantitative, PCR on the WAVE® nucleic acid fragment analysis system, Application Note 105, Tansgenomic, 1999.
3. C. G. Huber, P. J. Oefner, and G. K. Bonn, Rapid analysis of biopolymers on modified non-porous polystryrene – divinylbenzene particles. Chromatographia, 37, 653, 1993.
4. G. Bonn, C. Huber, and P. Oefner, Nucleic acid separation on alkylated nonporous polymer beads, U. S. Pat. 5,585,236, 1996.
5. C. G. Huber, P. J. Oefner, E. Preuss, and G. K. Bonn, High-resolution liquid chromatography of DNA fragments on highly cross-linked poly(styrene-divinylbenzene) particles, Nuceic Acids Res. 21, 1061, 1993.
6. P. J. Oefner and P. A. Underhill, Detection of nucleic acid heteroduplex molecules by denaturing high-performance liquid chromatography and methods for comparative sequencing, U. S. Patent 5,795,976 , 1995.
7. P. J. Oefner and P. A. Underhill, DNA Mutation detection using denaturing high-performance liquid chromatography, in N. C. Dracopoli, J. L. Haines, B. R. Korf, D. T. Moir, C. C. Morton, C. E. Seidman, J. g. Seidman, D. R. Smith, Eds, Current Protocols in Human Genetics, Wiley & Sons: New York, p 7.10.1 – 7.10.12, 1998.
8. P. A. Underhill, L. Jin, A. A. Lin, S. Q. Mehdi, T. Jenkins, D. Vollrath, R. W. Davis, L. L. Cavalli-Sforza, P. J. and Oefner, Detection of numerous Y chromosome biallelic polymorphisms by denaturing high performance liquid chromatography. Genome Res. 7, 996, 1997.
9. P. A. Underhill, L. Jin, R. Zemans, P. J. Oefner, and L. L. and Cavalli-Sforza, A pre-columbian Y chromosome-specific transition and its implications for human evolutionary history. Proc. Natl. Acad. Sci. 93, 196, 1996.
10. J. A. Thompson, A review of high performance liquid chromatography in nucleic acids research I. Historical perspectives, Biochromatography, 1, 16, 1986.
11. G. Zon and J. A. Thompson, A review of high performance liquid chromatography in nucleic acids research II. Isolation, purification, and analysis of oligodeoxyribonucleotides, Biochromatography, 1, 22, 1986.
12. J. A. Thompson, A review of high performance liquid chromatography in nucleic acids research III. Isolation purification, and analysis of supercoiled plasmid DNA, Biochromatography, 1, 68, 1986.
13. J. A. Thompson, A review of high performance liquid chromatography in nucleic acids research IV. Isolation, purification, and analysis of DNA restriction fragments, Biochromatography, 2, 4, 1987.
14. J. A. Thompson, A review of high performance liquid chromatography in nucleic acids research V. Nucleic acid affinity techniques in DNA and RNA research, Biochromatography, 2, 68, 1987.
15. J. A. Thompson, S. Garfinkel, R. B. Cohen, and B. Safer, A review of high performance

liquid chromatography in nucleic acids research VI. Nucleic acid affinity techniques in DNA-binding Protein research, Biochromatography, 2, 166, 1987.

16. J. A. Thompson, R. W. Blakesley, K. Doran, C. J. Hough, and R. D. Wells, Purification of nucleic acids, by RPC-5 analog chromatography: peristaltic and gravity-flow application. Methods, in Enzymology, Vol. 100 Academic Press: New York, pp 368–399, 1983.
17. A. P. Green, J. Burzynski, N. M. Helveston, G. M. Prior, W. H. Wunner, and J. A. Thompson, HPLC purification of synthetic oligodeoxyribonucleotides containing base- and backbone- modified sequences, Biotechniques, 19, 836, 1995.
18. A. P. Green, G. M. Prior, N. M. Helveston, B. E. Taittinger, X. Liu, and J. A. Thompson, Preparative purification of supercoiled plasmid DNA for therapeutic applications, Biopharm, pp 52–62, May 1997.
19. H. Ellegren and T. Laas, Size-exclusion chromatography of DNA restriction fragment, Fragment length determinations and a comparison with the behavior of proteins in size-exclusion chromatography, J. Chromatogr. 467, 217, 1989.
20. G. Bernardi, Chromatography of nucleic acids on hydroxyapatite column in L. Grossman and K. Moldave eds., Methods in Enymology, Vol. 21, Academic Press: New York, pp 92–140, 1971.
21. M. Soyab and Y. Sen, The isolation of extrachromosomal DNA by hydroxyapatite chromatography in R. Wu ed., Methods in Enzmology, Vol. 68, Academic Press: New York, pp 199–206, 1979.
22. A. D. Kelmers, G. D. Novelli, and M. P. Stulberg, Separation of transfer ribonucleic acids by reverse phase chromatography, J. Biol. Chem., 240 3979, 1965.
23. A. D. Kelmers, H. O. Weeren, J. F. Weiss, P. S. Pearson, M. P. Stulber, and G. d. Novelli, reversed-phase chromatography systems fro transfer ribonucliec acid – preparatory scale methods, in Methods in Enzymology, K. Moldave and L. Grossman, eds., Vol. 20, Part C, Academic Press: New York, pp 9–34, 1971.
24. R. L. Pearson, J. F.Weiss and A. D. Kelmers, Improved separation of transfer RNA"s on polychlorotrifluoroethylene-supported reversed-phase chromatography columns, Biochem. Biophys, Acta, 228, 770, 1971.
25. R. D. Wells, S. C. Hardies, G. T. Horn, B. Klein, J. E. Larson, S. K. Neuendorf, N. Panayatatos, R. K. Patient, and E. Selsign, PRC-5 column for the islation of DNA fragments, in Methods in Enzymology, L. Grosssman and K. Moldave eds., Vol 65, Academic Press: New York, pp 327–347, 1980.
26. J. A. Thompson and R. W. Blakesley, Higher performance liquid chromatography purification of nucleic acids, Fed. Proc. 42, 1953, 1983.
27. R. D. Wells, High performance liquid chromatography of DNA, J. Chromatogr., 336, 3, 1984.
28. R. A. Harwick and P. R. Brown, The performance of microparticle chemically-bonded anion-exchange resins in the analysis of nucleotides, J. Chromatogr., 112, 650, 1975.
29. E. Westman, S. Eriksson, T. Laas, P. Pernemalm, and S. E. Skold, Separation of DNA restriction fragments by ion-exchange chromatography on FPLC columns Mono P and Mono Q, Anal., Biochem., 166, 158, 1987.
30. F. Fack and V. Sarantoglou, Curved DNA fragments display retarded elution upon anion exchange HPLC, Nucleic Acids, Res., 19, 4181, 1991.
31. W. Bloch, Precision and accuracy of anion-exchange separation of nucleic acids, U. S. Pat. 5,856,192, 1999.
32. E. D. Katz and M. W. Dong, Rapid analysis and purification of polymerase chain reaction products by high-performance liquid chromatography, Biotechniques,8, 546, 1990.
33. E. D. Katz, Protocol: Quantitation and purification of polymerase chain reaction products by high-performance liquid chromatography, Molecular Biotechnology, 6, 79, 1996.
34. E. D. Katz, L. A. Haff, and R. Eksteen, Rapid separation, quantitation and purification of products of polymerase chain reaction by liquid chromatography, J. Chromatogr. 512, 433, 1990.
35. J. M. Wages, X. Zhao, and E. D. Katz, High-performance liquid chromatography analysis of PCR products, PCR Strategies, Academic Press: San Diego , pp 140–153, 1995.
36. S. Eriksson, G. Glad, P Pernemalm, and E. Westman, Separation of DNA restriction fragments by ion-pair chromatography, J. Chromatogr., 359, 265, 1986.
37. M. Kwiatkowski, A. Sandstrom, N. Balgobin, and J. Chattopadhyaya, J. Acta Chem. Scand., Series B: Org. Chem. and Biochem.,

Use of reverse phase ion pair chromatography to fractionate and purify DNA fragments and monomeric components of RNA, 38, 721, 1984.
38. W. Jost, K. Unger, and G. Schill, Reverse-phase ion-pair chromatography of polyvalent ions using oligonucleotides as model substances, Anal. Biochem., 119, 214, 1982.
39. D.A. Usher, Reverse-phase HPLC of DNA restriction fragments and ribooligonucleotides on uncoated Kel-F powder, Nucleic Acids. Res., 6, 2289, 1979.
40. C.G. Huber and G.N. Berti. Detection of partial denaturation in AT-rich DNA fragments by ion-pair reversed-phase chromatography, Anal. Chem., 68, 2959 1996.
41. C.G. Huber, P.J. Oefner and G.K. Bonn, Rapid and accurate sizing of DNA fragments by ion-pair chromatography on alkylated nonporous poly(styrene-divinylbenzene), Anal. Chem., 67, 578 1995.
42. P.J. Oefner and G.K. Bonn, High-resolution liquid chromatography of nucleic acids, Am. Lab., 26, 28C 1994.
43. P.J. Oefner, C.G. Huber, F. Umlauft, G.N. Berti, E. Stimpfl, and G.K. Bonn, High-resolution liquid chromatography of fluorescent dye-labeled nucleic acids, Anal. Biochemi., 223, 39,1994.
44. A. Hayward-Lester, P.J. Oefner, S. Sabatini, and P.A. Doris, Accurate and absolute measurement of gene expression by single tube RT-PCR and HPLC, Genome Research, 5, 494, 1995.
45. P.A. Doris, B.S. Chilton, P.J. Oefner, and A. Hayward-Lester, Quantitative analysis of gene expression by ion-pair high-performance liquid chromatography, J. Chromatogr., 806, 47, 1998.
46. A. Hayward-Lester, P.J. Oefner, S. Sabatini, D.B. Kainer, C.A. Hinojos, and P.A. Doris, Modeling and analysis of competitive RT-PCR. Nucleic Acids Res., 26, 2511, 1998.
47. K.H. Hecker, B. Turpie, and A. Kuklin, Optimization of cloning efficacy by pre-cloning DNA fragment analysis, BioTechniques, 26, 216, 1999.
48. A. Azarani, and K.H. Hecker, RNA analysis by Ion-Pair Reversed Phase High Performance Liquid Chromatography, Nucl. Acid Res., 2001, Vol. 29, No. 2 e7.
49. A. Azarani, R. Haefele, and K.H. Hecker, RNA analysis by ion-pair reversed-phase HPLC, Nature Biotechnology, 2000, Volume II, 29.
50. A. Hayward-Lester, P.J. Oefner, D. Kainer, C.A. Hinojos, and P.A. Doris, Kinetics of competitive RT-PCR, in: PCR Methods Manual, M. Innis, D. Gelfand and J. Sninsky, Eds., Academic Press, San Diego, 1998, pp 231–261.
51. M.A. Marino, J.M. Devaney, J.K. Smith and J.E. Girard, Sequencing using capillary electrophoresis of short tandem repeat alleles separated and purified by high performance liquid chromatography, Electrophoresis, 19, 108, 1998.
52. P.A. Underhill, P. Shen, A.A. Lin, L. Jin, G. Passarino, W.H. Yang, E. Kauffman, B. Bonne-Tamir, J. Bertranpetit, P. Francalacci, M. Ibrahim, T. Jenkins, J.R. Kidd, S.Q. Mehdi, M.T. Seielstad, R.S. Wells, A. Piazza, R.W. Davis, M.W. Feldman, L.L. Cavalli-Sforza, and P.J. Oefner, Y chromosome sequence variation and the history of human populations, Nat. Genet., 26, 358, 2000.
53. B. Hoogendoorn, M.J Owen, P.J. Oefner, N. Williams, J. Austin, and M. O"Donovan, Genotyping single nucleotide polymorphisms by primer extension and high performance liquid chromatography, Human Genetics, 104, 89, 1999.
54. K. Kosaki, H. Yoshihashi, Y. Ohashi, R. Kosaki, T. Suzuki, and N. Matsuo. Fluorescence-based DHPLC for allelic quantification by single-nucleotide primer extension, J. Biochem. Biophys. Methods, 47, 111, 2001.
55. J.M. Devaney, E.L. Pettit, S.G. Kaler, P.M. Vallone, J.M. Butler, and M.A. Marino, Genotyping of two mutations in the HFE gene using single-base extension and high-performance liquid chromatography, Anal. Chem., 73, 620, 2001.
56. B. Hoogendoorn, N. Norton, G. Kirov, N. Williams, M. Hamshere, G. Spurlock, J. Austin, M. Stephens, P. Buckland, and M. Owen, Cheap, accurate and rapid allele frequency estimation of single nucleotide polymorphisms by primer extension and DHPLC in DNA pools, Hum. Genet., 107, 488, 2000.

2
Instrumentation and Operation

2.1
Introduction

The name "DNA Chromatography" applies to any modern HPLC (high performance liquid chromatography) method for chromatographic separation of DNA. Although HPLC equipment and its use are well established for analytical work, the use of HPLC for DNA separation and purification is more recent and not as well established. Nevertheless, more than 30 years of instrumentation development and refinement and an even longer time devoted to research to understand fundamental chromatographic principles have allowed the development of a robust and powerful technology for nucleic acid analysis and purification. The hardware is rugged and the software is well advanced.

The running joke in graduate laboratories of the first HPLC chromatographers was that because of poor fittings, leaks, temperamental pumps, and the constant care and feeding that was needed to keep a HPLC operating, the chromatographer had to be more of a plumber than a scientist. (This actually was somewhat of a comforting thought to the graduate student since it was known that plumbers made quite good money. Perhaps the training could be put to good use just in case the student's graduate advisor ever suggested that perhaps plumbing would in fact be a more suitable profession.) Since those early days, HPLC equipment has become much more refined. The improved engineering and software has resulted in longer lasting seals, much more reproducible operation and fewer breakdowns. Because of these refinements, the operation of DNA Chromatography instrumentation is usually automated and trouble free.

It is not necessary for the reader to have had previous experience with HPLC instrumentation or chromatographic principles in order to use and understand this book. However, references are included at the end of this chapter, and after reading this chapter the reader may wish to explore the topic further. However, it should be noted that DNA Chromatography has special instrumentation, software, chemical, and molecular biological components which are necessary for the successful practice of the technology – the discussion that is in this book. It should also be noted that with modern software and hardware technology, the technique of DNA Chromatography can be practiced without knowing all details of the technology. Instrumentation can operate automatically, and software can facilitate the process of analysis and purification. Nevertheless, understanding the basic principles of a particular technology will help the scientist make full use of the technology. This is especially important in DNA Chromatography where the authors believe the full potential of the technique has yet to be realized and practiced.

2.2
General Description of the DNA Chromatograph

Figure 2.1 shows a block diagram of the general components of a DNA Chromatography instrument. The components are: a supply of eluents or buffers (also called the mobile phase), an eluent degasser to remove dissolved oxygen, carbon dioxide and nitrogen from the fluid, a high pressure pump (with pressure indicator) to deliver the eluent or mobile phase, an autosampler including syringe and sample valve for introducing sample into the eluent stream and onto the column, a column (also called the stationary phase) to separate the sample mixture into the individual components, an oven to contain the column and control the temperature of the nucleic acids, a detector to measure the peaks or bands of the nucleic acid fragments as they elute from the column, an optional fragment collector to collect any particular fragment of interest and a data system for collecting and organizing the chromatograms. A computer controls the instrument. Software is used to develop and implement analytical methods for each set of samples. The software guides the user to calibrate the instrument and to develop the appropriate

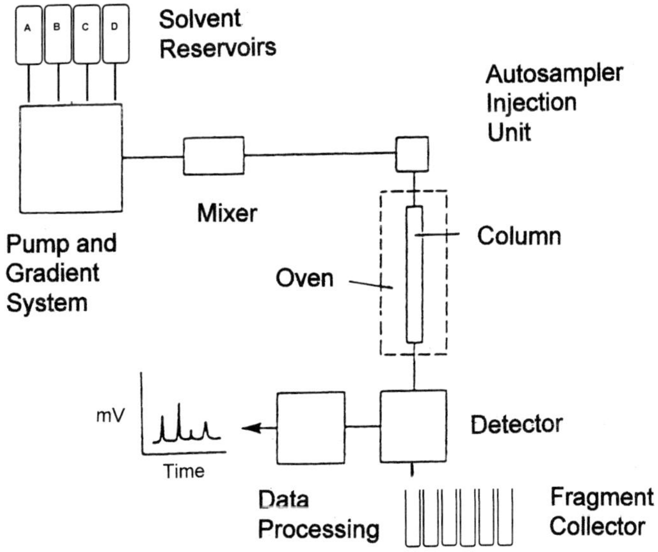

Figure 2.1. Block diagram of DNA Chromatograph.

and optimum methods depending on the analysis being performed and whether fragments are to be collected.

A key component of the instrument is the separation column. Unlike electrophoresis gels that are used only once, the chromatographic column is used for several hundred or even thousands of injections before it is replaced. The eluent is pumped continuously through the column. Periodically, samples are injected and separated. A particular mix of eluent strength is used for each separation depending on the sample type (fragment sizes) and the desired separation.

A pump is used to force the eluent or buffer through the system, including the column, at a fixed rate, such as 0.9 ml/min. In the sample FILL mode a small volume of sample (typically 4 μL) is pulled into the autosampler syringe and placed into the sample loop. At the same time eluent is being pumped through the rest of the system, while bypassing the sample loop. In the sample INJECT mode, a valve is turned so that the eluent sweeps the sample from the sample loop into the column. A detector cell of low dead volume is placed in the system just after the column. The detector is connected to a data-acquisition device so that a chromatogram of the separation (signal vs. time) can be plotted automatically. A UV-visible detector is most often used in DNA Chromatography. Because DNA absorbs UV light, a signal results when the fragment band travels past the detection window. This absorption of light is detected and amplified and recorded as a function of time to give a chromatogram. Finally, a fragment collector can be positioned at the outlet tube of the detector. When a fragment of interest is detected the deposition probe can be directed to a collection vial. After the material has been collected the probe is directed back to waste.

2.3 Detailed Description of the DNA Chromatograph

2.3.1 The General Instrument and Materials

This section is a detailed description of the various components of a DNA Chromatography instrument, their function, and some general points for upkeep of the chromatograph. New users can use the information not only to operate an instrument, but also to understand how an instrument is built and to recognize the parts of the instrument that may need maintenance. The hardware is similar to that used for high pressure liquid chromatography but does have some important differences. Readers who are familiar with HPLC will recognize the similarity and the differences to the DNA Chromatograph.

The reader should again refer back to Figure 2.1. Fluid is pumped through the system under high pressure. Some of the back pressure is due to the small diameter tubing, but most of the pressure is due to the separation column that is located in the oven. Everything on the high pressure side, from the pump outlet to the end of the column, must be strong enough to withstand the pressures involved. The wetted parts are usually made of stainless steel, titanium, PEEK (polyetheretherketone) and other types of plastics. Materials, such as sapphire, ruby, or even ceramics are used in the pump heads, check valves, and injector of the system. PEEK and titanium are the materials of choice for the DNA Chromatograph. Stainless steel is also a material of choice provided the system is properly conditioned to remove and control internal corrosion. Stainless steel components are considered to be more reliable than those made from plastics, but require more care. The stainless steel chromatograph is normally delivered from the manufacturer pretreated so that corrosion is not present. The reader is advised to consult the instrument manufacturer for care and upkeep instructions. More discussion on the effect of instrument corrosion is included in Chapter 3.

2.3.2 Dead Volume

Dead volume is any empty space or unoccupied volume. The dead volume of a chromatographic system is between the point where the sample is introduced (the injector) to the point where the peak is detected (the detection cell) and must be kept to a minimum. The presence of too much dead volume can lead to extreme losses in separation efficiency due to broadening of the peaks. Although all regions in the flow path are important, the most important region where peak broadening can happen is in the tubing and connections from the top end of the separation column to the detector cell. It is important to follow the manufacturer's instructions when changing the column for making other fluid connections. If fragments are to be collected, the dead volume from the end of the detector cell to the end of the deposition probe is also extremely important to maintain sharp and crisp fragment collections with no cross contamination of neighboring peaks.

Of course, there is a natural amount of dead volume in a system due to the internal volume of the connecting tubing, the interstitial spaces between the column packing beads and so on. But using small bore tubing (0.007 inch, 0.18 mm or smaller) in short lengths when making the injection to the column and the column to detector connections is important. Also, it is important to make sure that the tubing end does not leave a space in the fitting when the connections are made. In general, the tubing should be butted against the bottom of the fitting first and then the screws of the fitting tightened (see Figure 2.2).

Dead volume before the pump and from the pump to the injector should also be reduced to facilitate rapid changes in the eluent composition in gradient elution. A certain amount of dead volume is needed to ensure complete mixing of the different eluent components before they reach the column. Too much dead volume will lengthen the time needed to change eluent concentration that is part of the gradient process, thus leading to longer separation times. A discussion of gradient formation is found later in this chapter. Discussions of the use of eluent gradients are found in the Chapter 3.

Eluent entering the pump and even more importantly entering the column should not contain any dust or other particulate matter. Particulates can interfere

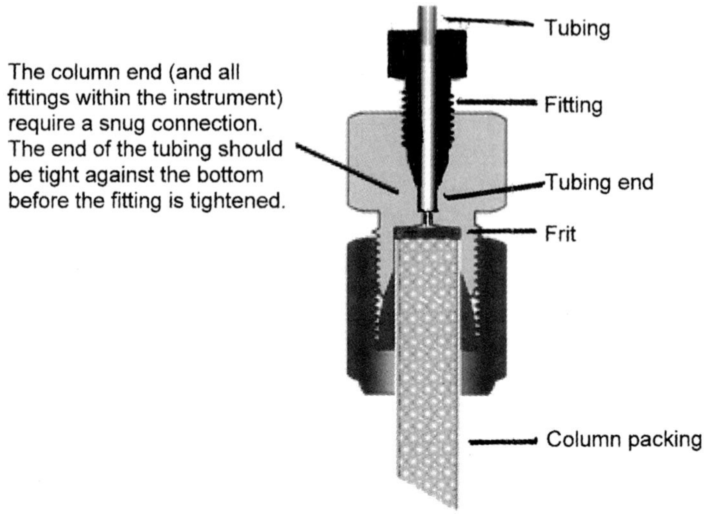

Figure 2.2 Column end and connection fitting with tubing. Note that the dead volume is minimized by making sure that the end of the tubing is butted down against the fitting.

with pumping action and damage the seal or valves. Material can also collect on the inlet frits or on the inlet of the column causing pressure buildup and premature column replacement. The components used to prepare the eluents are normally filtered using a 0.2 or 0.45 micron nylon filter. Filters are made of other materials; the user should determine if the material is compatible with the solvent before use. Note after the eluent is prepared, it is normally not filtered. Because acetonitrile solvent is quite volatile, filtering can remove or change its concentration in the eluent. This would change the eluting strength of the eluent. For this reason, glassware and other containers used to prepare and store the eluent should be rinsed with particulate free solvent and dried carefully before use. (Again, there should be no residual solvent in the containers that would change the concentration of the eluent).

2.3.3
Degassing the Eluent

Degassing the eluent is important because air can get trapped in the pump check valves (discussed later in this section) causing the pump to lose its prime. Loss of prime results in erratic eluent flow or no flow at all. Sometimes only one pump head will lose its prime and the pressure will fluctuate in rhythm with the pump stroke. Another reason for removing dissolved air from the eluent is because oxygen in the fluid can cause it to become highly corrosive. Removal of the oxygen through degassing will result in less required maintenance of the stainless steel components of the chromatograph.

Usually degassed water is used to prepare eluents and efforts should be made to keep exposure of eluent to air to a minimum after preparation. The best way to remove gases is using an inline degasser. Modern inline degassers are becoming quite popular. These are small devices that contain 2–4 channels and the eluent travels through these devices from the eluent reservoirs to the pump. The tubing in the device is gas permeable. The tube is surrounded by vacuum which causes gases in the fluid to be transported through the tubing wall to the vacuum and leaves the eluent ions and higher boiling fluids behind.

It should be noted that when the pump flow is stopped or on standby, this will cause the eluent to have a very long residence time in the degasser. The organic solvent (acetonitrile) in the eluent will gradually go through the membrane and change the concentration of the eluent. Therefore, it is best to flush this fluid out of the instrument when it is first started up. The best method of accomplishing this is to perform a blank gradient run. This is a run where no sample is injected, but the pump goes through the gradient process. Failure to do this would result in the first real run to be erratic. Finally, it is best to change the eluents every day (or at least every couple of days) to keep the concentration accurate and also to prevent bacterial growth in the reservoirs. If bacteria are pumped through the system, they are likely to build up at the top of the column and increase the back pressure of the system, sometimes quickly and dramatically and at other times gradually.

2.3.4 Pumps

Chromatographic pumps are designed around an eccentric cam that is connected to a piston (Figure 2.3). The rotation of the motor is transferred into the reciprocal movement of the piston. A pair of check valves controls the direction of flow through the pump head (discussed below). A pump seal surrounding the piston body keeps the eluent from leaking out of the pump head. A piston seal leak is usually formed at the back side of the pump head.

In single-headed reciprocating pumps, the eluent is being delivered to the column for only half of the pumping cycle. A pulse dampener is used to soften the pulsating action of pressure of the pumping cycle and to provide an eluent flow when the pump is refilling. Use of a dual head pump is better because heads are operated 180° out of phase with each other. One pump head is pumping while the other is filling and vice versa. The eluent flow rate is usually controlled by the pump motor speed although there are a few pumps that control flow rate by control of the piston stroke distance.

Figure 2.4 shows how the check valve works. On the intake stroke, the piston is withdrawn into the pump head causing a suction. The suction causes the outlet check valve ball to settle onto its seat and prevent back flow to occur. At the same time the inlet check valve ball rises from its seat allowing eluent to fill the pump head. Then the piston travels back into the pump head on the delivery stroke. The pressure increase of the eluent seats the inlet check valve ball preventing back flow. At the same time the outlet valve is opened. The eluent is forced out

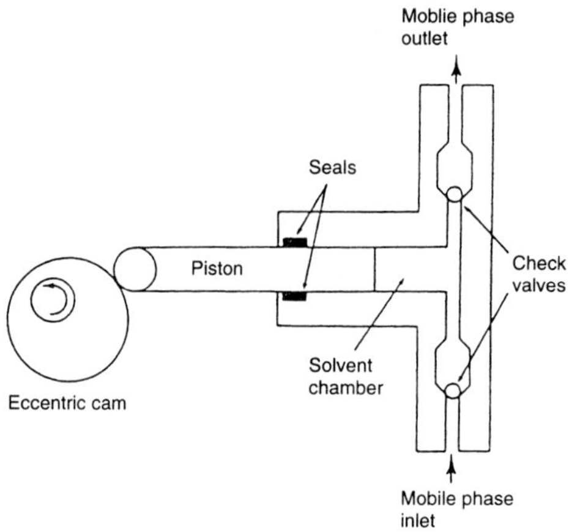

Figure 2.3 DNA Chromatograph cam, pump head, piston, piston seals and check valves.

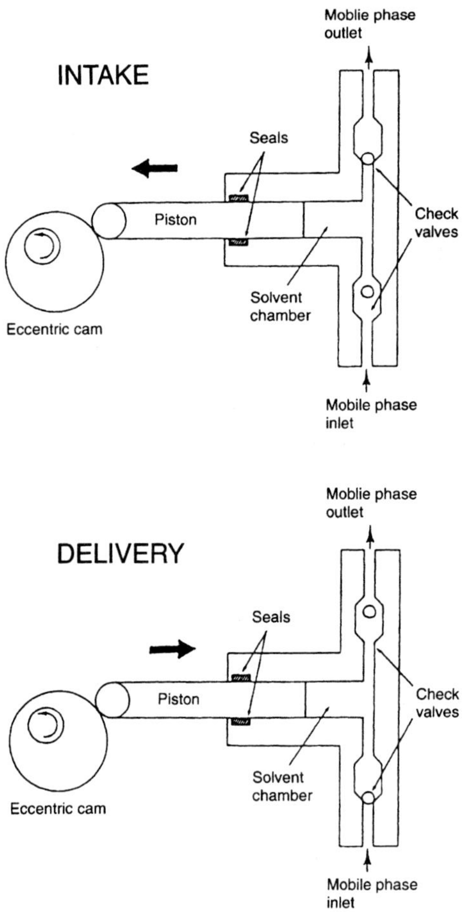

Figure 2.4. Check valve positions during intake and delivery strokes of the pump head piston.

of the pump head through the outlet check valve and through the autosampler injection valve and then the column. Failure of either of the check valve balls to seat properly will cause pump head failure and eluent will not be pumped. In most cases, this is due to air being trapped in the check valve so that the ball cannot seat properly. Flushing or purging the head usually will take care of this problem. Using degassed eluents is also helpful. In a few cases, particulate material can prevent seating of the check valve. In these cases, the check valve must be cleaned by flushing or replaced. A common cleaning solvent is dilute nitric acid. The instrument manufacturer has instructions on how to perform this operation.

2.3.5
Gradient Formation

Isocratic separations are performed using an eluent at a constant or uniform concentration of eluent solution. While it is desirable (simpler) to perform nucleic acid separations with single isocratic eluent, it is almost always necessary to form a gradient of the eluent strength. A gradient is weak eluent first entering the separation column and then gradually becoming a concentrated, strong eluent over the course of a chromatographic run. This allows the separation of nucleic acids that may have a wide range of affinities for the column. Weakly sticking nucleic acid fragments elute first and then as the eluent concentration is increased, more strongly sticking nucleic acid fragments can be eluted by the stronger eluent.

There is the old adage that a weakness of something can become its strength if only used in the correct way. When compared to electrophoresis where gradients are rarely used, gradients in DNA Chromatography add a complexity for fragment separation. But this is also one of the major strengths of DNA Chromatography. The concentration of acetonitrile that is present in the column at any particular instant dictates the size of the fragment that is being separated by the column. The separation conditions are controlled in "real time" or instantly so that targeted materials can be eluted and collected as desired. Smaller fragments require a lower concentration of acetonitrile and larger fragments require a higher concentration. The range and the duration of the gradient determines how fast the separation will be and what fragment size range will be measured. This feature can be changed from one separation run to the next at will. Control of the hardware components of gradient formation is discussed.

Figures 2.5 and 2.6 show the two most popular methods for forming gradients. The first method is called high pressure gradient mixing. Flow from two high pressure pumps is directed into a high pressure mixing chamber. One pump contains a weak eluent while the other contains the stronger eluent. After the mixing chamber, the flow is directed to the injector and then onto the column. Controlling the relative pumping rate delivery of each pump forms the gradient. The total combined flow from the two pumps is held constant. A high flow of the weak eluent pump and a low flow of the strong eluent pump initiates the gradient. Then over the course of the chromatographic run, the relative flow rate of the strong eluent pump is increased while the flow rate of the weak eluent pump is decreased keeping the total flow rate constant.

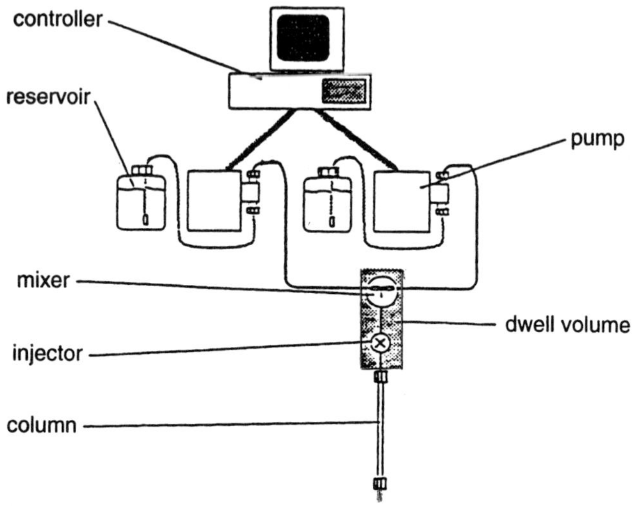

Figure 2.5. High-pressure mixing systems use two or more independent pumps to generate the gradient. Advantages of high pressure mixing are smaller dwell volume and faster gradient formation.

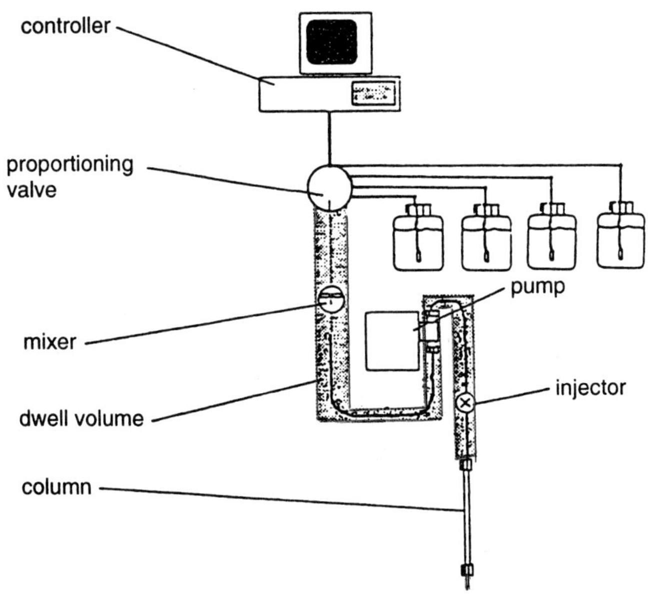

Figure 2.6. Low-pressure mixing systems use a single pump with a proportioning valve to control composition. Advantages of low pressure mixing are lower cost (single pump) and more versatile gradients (four solvents).

A more popular and more rugged method of forming gradients is low pressure gradient mixing. In this method, a single pump contains 3 or 4 micro proportioning valves at the inlet of the pump. At low pressure, gradients can be formed from eluents A, B, C, and D (or any combination) by metering controlled amounts from the various eluent reservoirs into the pump. The composition in the low pressure mixing chamber is controlled by using timed proportioning valves. Only one reservoir valve is open at any time. The total cycle time remains constant throughout the gradient, but the time that any one of the reservoir valves remains on will vary. At the start of the gradient, the valve connected to the weak eluent is open longest. As the gradient progresses, the valve connected to the strong eluent becomes open for a longer and longer time while the weaker eluent valve is open for shorter times. The time cycle of the valve remains constant. Up to four valves are often available to give options for different types of gradients or the use of cleaning solutions. But generally, gradients are formed with just two of the valves.

This method of gradient generation is less expensive than high pressure gradient formation because only one pump is used. It is usually more rugged because the extra pump and check valves are more likely to fail than micro switching valve of a low pressure gradient system. On the other hand, high pressure gradients can be faster because there is less dwell volume (gradient delay volume) in this gradient formation system resulting in a quicker gradient reaching the column and detector. The actual time of gradient formation depends in part on the volume of the mixer located after the eluent gradient has been first formed. Mixers for both types of gradient tend to have large volumes (to ensure complete mixing) so the time for the gradient to reach the column is usually not much different for the two types of systems. The gradient delay volume is different from the dead volume (see description of dead volume earlier and definitions in the glossary).

2.3.6 Pressure

Column inlet pressures can vary from 500 psi up to perhaps a high of 3,500 psi. with normal operating pressures around 1,500 psi. The pressure limit on a DNA Chromatograph is usually 4,000 to 5,000 psi depending on the fittings and other hardware used. The eluent back pressure is directly proportional to the eluent flow rate. Although still popular, psi (pounds per square inch) is gradually being replaced by more modern terms for pressure measurement i. e. 1 bar = 1 atm (atmospheres) = 14.5 psi = 10^5 Pa (Pascal).

2.3.7 Autosampler Injector

The injections are normally taken directly from a 96-well or 384-well PCR thermal cycler plate. It is possible to manually inject a sample, but this is rarely done. In any case, both manual and autosamplers rely on the injection valve. An injection valve is designed to introduce precise amounts of sample into the sample stream.

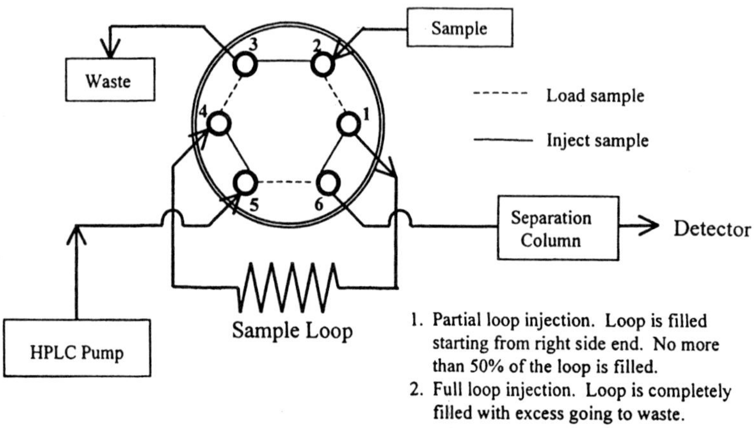

Figure 2.7. The injection valve is found in the autosampler. This is a schematic depicting partial and full loop injection methods. Partial loop injection is most common.

The variation of sample volume injected is usually less than 0.5 % from one injection to the next injection. Figure 2.7 shows a schematic of the valve. It is a 6 port and 2 position device where one position is load and the other is inject. In the load position, the sample is first cleaned with an external solution and then the syringe takes up sample from PCR vial and pushes the sample into the injection loop. (Note the external syringe needle cleaning solution must have an organic solvent concentration that is lower or equal to the starting gradient. This will prevent premature initiation of the gradient).

The loop may be partially filled (partial loop injection) or completely filled (full loop injection) (see Figure 2.7). Partial loop injection is by far the most popular due the small volume injections performed in DNA Chromatography. In partial loop injection most or all of the sample taken up by the syringe is ultimately injected into the system whereas in full loop injection much of the sample is lost. Partial loop injection depends on the precision filling of the loop with a small known amount of material. If partial loop injection is used, the loop must not be filled to more than 50 % of the total loop volume or the injection may not be precise. In full loop injection, the sample is pushed completely through the loop. Typical loop sizes are 10 – 200 μL. Normally at least a two fold amount of sample is used to fill the loop with excess sample from the loop going to waste.

At the same time that the sample loop is being loaded with sample, the eluent is traveling in the by-pass channel of the injection valve to the column. An injection of the sample is accomplished by turning the valve and placing the injection loop into the eluent stream. Usually the flow of the eluent is opposite to the flow of loading sample into the loop. The injected sample travels to the head of the column as a slug of fluid. The fragments in the sample interact or adsorb onto the column. The separation process is started with eluent of the appropriate strength pushing the sample components down the column. Injection valves require periodic maintenance and usually have to be serviced after about 5,000 injections. The manual for the instrument should be consulted for details on service.

2.3.8
Separation Column

The separation column (sometimes called a cartridge) is a small cylinder that contains packing material used to separate the nucleic acid fragments. The packing materials are held inside the column with porous frits that are located on each end of the column. The frits allow passage of the fluids entering and leaving the column, and hold the packing material in place inside the column.

Columns for chromatography are packed with beads that provide the basis for the separations. The column packing is normally performed by capping the bottom of the column with a frit and connecting the column to a reservoir. A slurry of the packing material and a solvent is prepared and poured into the reservoir. Then a pump is connected to the top of the reservoir and a fluid is pumped under high pressure forcing the slurry into the column. The column packing material is trapped in the column and eventually the column is filled. After a suitable time, the pump is stopped, the reservoir is disconnected and the top of the column is capped. The column is tested before use to ensure that the packing procedure was successful.

A short description of column materials is given here and the main discussion is given in Chapter 3. There are essentially two types of column materials that have been used for DNA separations. The most popular is a reverse phase (neutral charged surface) material. A popular trademark name for this material is called DNASep® column and is supplied by Transgenomic, Inc., San Jose, CA and Omaha, NE. Another, older type of column material is an anion exchanger. Although, most of the recent work has been performed with neutral reverse phase materials, anion exchange is still used, especially for the separation of short, single-stranded DNA.

The most popular column materials are polymer-based because they are rugged and can withstand extremes in eluent pH. There are also silica based column materials on which a neutral surface has been applied. The column is carefully packed with a spherical particle with a typical diameter of about 2 μm. Most column packings are functionalized with a hydrophobic, neutral C-18 alkyl group or they are functionalized with quaternary ammonium groups, which serve as the sites for an anion exchange process.

A typical column used in DNA Chromatography might be 50 mm long and a bed diameter of 4.6 mm although columns much shorter or longer can be used. Preparative columns are used to purify larger amounts of DNA fragments. These columns use larger bed diameters as the need for purifying material increases. Bed diameters of 7.8, 11 mm or even larger diameters are common.

Reusable PEEK fittings are used almost exclusively to connect tubing to columns and other instrument components. As stated earlier, the tubing should be bottomed out or pushed completely into the column end before the fitting is tightened to ensure that there is not unnecessary dead volume in the connection.

2.3.9 Column Protection

Column protection not only extends the useful life of the separation column, but proper protection of the separation column can also result in more reliable analytical results over the lifetime of the column. Scavenger columns, located between the pump and injector, are one means of protecting the column. The scavenger removes particulate material that may be present in the eluent. The scavenger may also contain a resin to "polish" the eluent of any dissolved contaminant. An example is a chelating resin to remove metal ion or colloidal contaminants. Besides protecting the separation column, scavenger columns may also improve detection of the fragments by reducing the detector background signal due to residual contaminants.

Another method of column protection is the use of the guard column or guard cartridge. The guard is located directly in front of the column and contains the same or similar packing as the main separation column. The frits located at the ends of the column are an efficient means to trap particulate materials including denatured proteins, colloidal metals, dust, etc. Dissolved contaminants are material in the sample that adsorb to the column packing and are not easily removed by cleaning procedures. Particulate or dissolved contaminants that would normally be trapped by the column are instead trapped on the guard column. The guard is less expensive than the main separation column and can be replaced more often. The main disadvantage of the guard is that it can affect the selectivity or retention times of the separation.

Most DNA Chromatographic users do not use scavenger columns or guards but rather prefer the use of inline filters located directly in front of the separation column. Material that would normally be trapped and contaminate the separation column instead gets trapped by the filter. In line filters are quite inexpensive and should be changed frequently – at least every 2 to 3 weeks and sometimes much more often. Filters do not affect the retention of nucleic acid fragments.

Another less common but useful approach for extending the life of the separation column is to reverse the direction of flow through the column. When the column is reversed, particulate or adsorbed material that have accumulated at the top of the column in effect becomes material at the bottom of the column and can be washed off with normal operation of the system. Reversal of the column can be done every 2 – 4 weeks over the life of the column.

Finally, the column should be cleaned with cleaning eluent throughout the lifetime of the column. Cleaning procedures must be implemented at the end of each run or after a set of injection. The reason this is important is that the concentration of the (acetonitrile) solvent used in the eluent during normal operation is quite low and rarely is above 25 %. While this gives useful separation of nucleic acids, contaminants usually need much higher concentrations of acetonitrile to wash them off the neutral column packing surface. Instrument manufacturers take this into account by providing a cleaning solvent to be used at the end of each run or after a set of runs has been completed. Failure to keep the column clean is probably the most common error that the user can make. For this reason, standards with known peak patterns should be injected at the beginning and end of a series of runs to ensure that the column is clean and working properly throughout the analysis of the samples.

2.3.10
Column Oven

The column oven is a valuable part of any DNA Chromatography instrument. There are some minor reasons for its use. Heating the eluent before it flows through the column lowers the viscosity of the eluent and consequently lowers the back pressure that the instrument must pump against. Since the column normally has a very high back pressure this reduces the strain on the fittings and seals.

Another benefit of using a column oven is improved detection of the DNA fragments. UV detection can be affected by changes in the refractive index of the fluid entering the detection cell. These changes contribute to background detector signal noise. The background noise caused by refractive index changes is usually small but can still affect the detection limits when very low concentration of materials are being measured. By controlling the fluid temperature entering the detection cell, the noise due to refractive index changes is minimized.

But the most compelling reason to use a column oven is to control the structure of the DNA. By controlling the temperature of the oven, separation conditions can be characterized as double-stranded (non-denaturing), DHPLC (partial denaturing), and single-stranded (full denaturing) conditions. DHPLC is partial denaturing HPLC for double-stranded detection of mismatched heteroduplex species and will be discussed in detail in Chapter 4. Full denaturing conditions will control or at least modify secondary structure that might be present in single-stranded nucleic acids. Secondary structures could affect the retention of nucleic acid resulting in multiple or changed retention.

Temperature should be thought of as a reagent in DNA Chromatography. Control of this reagent will help in designing experiments to detect mutations. Virtually all ovens that are used for normal HPLC operation are unsuitable for DNA Chromatography. First, it is not the control of the oven that is important, but it is the control of the eluent fluid entering the column that is important. Because the fluid entering the oven compartment is cooler than the oven, there is a time lag before the fluid reaches the oven temperature. In most HPLC ovens the fluid never does reach the set point of the oven. Figure 2.8 shows a preheat coil that is positioned ahead of the column. As the fluid goes through this coil, the eluent (and injected sample) gradually heats up so that when it enters the column it has been heated to an accurate temperature. Also, the oven temperature should not drift and should be precise i. e. come to the same temperature each time it is directed by the run method to go to a specific temperature.

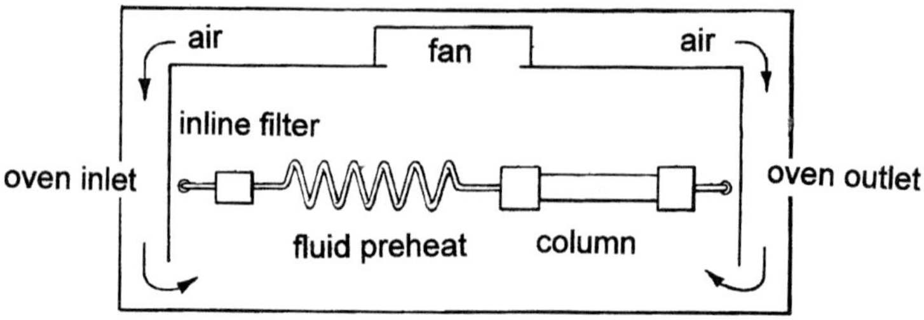

Figure 2.8. Oven compartment showing preheat coil, inline filter and placement of the separation column. The eluent must heat up to the oven set point before entering the column. (From Ref. [1] with permission).

The oven remains one of the most critical and difficult parameter to control in the DNA Chromatograph. The reader should consult the manufacturer for information on oven use, calibration and upkeep.

2.3.11 Detection

2.3.11.1 Selective vs. General Detection

As the name implies, general detectors will detect all nucleic acids. The UV detector can be said to be a general detector for nucleic acid detection. Selective detectors will detect only fragments that have a certain detectable property. In the case of fluorescence, only fragments that have been tagged with a fluorescent molecule will be detected. (An exception to this which can make fluorescence detection general will be discussed later). Selective detector can be extremely effective in picking out a single fragment of interest from a high eluent background or a high sample matrix background.

Another key advantage to selective detection is the possibility of achieving lower detection limits. A selective detector may or may not have a higher sensitivity (signal per unit concentration) for a particular ion. However, the lower background signals produced by selective detection will translate into lower detection limits because the signal to noise ratio is improved.

It is important to review some definitions. Detector sensitivity is the amount of signal that is recorded per unit of sample concentration. The detection limit depends on the sensitivity and the detector noise. The detection limits definition used by analytical chemists is the amount of sample that is detected at a signal that is 3 times the detector noise level. However, practical detection limits are when the signal is 10 times the detector noise level. Molecular biology sensitivity means something else. For example the mutation detection sensitivity is based on the use of mutation standards or controls. If a particular control can be detected by a method, then it is said that the mutation is sensitive to detection by that method.

Certainly fluorescence is the detection method that first comes to mind for nucleic acid analysis. It is the detection method of choice for electrophoresis separations. Four color fluorescence detection probes are used for sequencing with each color corresponding to a particular nucleotide. It is also true that it is difficult to use UV detection for electrophoresis detection because of interferences and poor selectivity and sensitivity. One of the great strengths of DNA Chromatography over capillary and slab gel electrophoresis is that UV detection can be used without any previous modification of the nucleic acid. No tagging is necessary for separation and purification of nucleic acids. Still, fluorescence has its strengths. And for general R&D purposes, many researchers use UV and fluorescence detectors simultaneously. They are connected in series with UV detection first followed by fluorescence detection. But if only one detector were available for an instrument, the general UV detector is the detector of choice.

2.3.11.2 **Ultraviolet-Visible Detectors**

Detection of nucleic acids is performed by UV spectrophotometric detection. It is important to have a basic understanding how a UV detector operates in order obtain the best performance. These detectors are sometimes called UV–VIS detectors or UV detectors for short.

Nucleic acids absorb light strongly in the UV with its maximum wavelength at 260 nm. Variable wavelength detectors are set at 260 nm for detection; however, single wavelength detectors work very well at 254 nm. In this case, an eluent has been selected to not absorb at these wavelengths. In some cases where there is absorption of the eluent due to contamination or decomposition, the baseline stability can be quite poor or the baseline will shift when a gradient run is performed. There is always a natural baseline shift when a gradient is performed, but too great of a shift, or an unreproducible baseline will render the instrument unusable.

The detector must be able to "pick out or see" nucleic fragments in the presence of the various components of the eluent molecules and solvent. Peaks containing as little as 0.3 ng of material are detectable, but usually one would prefer to work with amounts 10 times higher. A UV detector will respond to all of the nucleic acid fragments that pass through the detector cell whether they are single-stranded DNA, various analogs, double-stranded DNA or RNA. Early researchers performing work on fragment collection questioned whether material collected after passing through a cell with UV light would be damaged. Would the damage prevent further processing of the DNA such as cloning or PCR? To date, there has been no data showing that UV detection will damage nucleic acid fragments that are collected.

Figure 2.9 shows the schematic by which the cell detects the DNA fragments. Light is directed through a window in a "Z" cell and a transducer measuring the light is positioned on the other end of the cell. The light that reaches the detector is measured continuously as a function of time. This measurement is shown as the baseline in the chromatographic separations. As the fragments travel through the

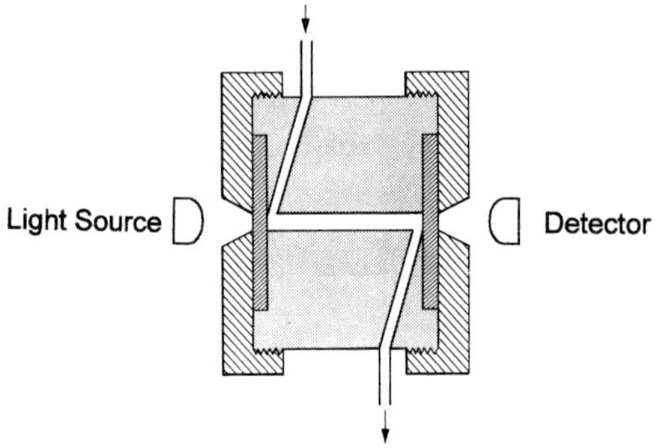

Figure 2.9. Schematic of "Z" flow cell for a UV detector.

cell, the fragments absorb light so that less light reaches the detector transducer. This reduction is amplified electronically which results in peaks in a chromatogram. As the amount of fragment material is increased, the amount of light reaching the detector transducer is reduced making larger or higher chromatographic peaks.

The fundamental law under which ultraviolet-visible (UV-VIS) detectors operate is the Lambert-Beer law. It can be stated in the following form:

$$A = \varepsilon \, b \, C \tag{2.1}$$

A is the absorbance of a species of concentration, C, that has an molar absorptivity ε, in a cell of length b. Concentration is usually expressed in molar concentration and the path length measured in cm. The term, ε, has units that are the inverse of C and b. This leaves A dimensionless; it is usually described in terms of Absorbance Units. A detector set to a certain sensitivity, for example 0.16, is said to be set at 0.16 Absorbance Units Full Scale sensitivity (0.16 AUFS sensitivity). The AU and AUFS may also be expressed as mV (milli Volts) and mVFS. The eluent should have a low or zero absorptivity. If it has any signal at all, the detector is "zeroed" with eluent flowing through the cell so that is defined to have no absorbance or no signal.

Although detectors that operate at a single 254 nm wavelength are available, most manufacturers provide UV detectors that have variable wavelength capability. Any UV wavelength can be chosen for detection. They may also have VIS (visible) wavelength capability. The 190 to 320 nm wavelengths are UV and the visible range is 320 to about 680 nm. The visible wavelength region is rarely used and is unnecessary for DNA Chromatography. Likewise, it is unnecessary to use a variable wavelength detector since virtually all work is performed at 260 nm wavelength (or in the case of a single wavelength detector all work is performed at 254 nm).

2.3.11.3 **Fluorescence Detector**

Fluorescence detection is closely related to absorption detection. After molecules have absorbed radiant energy and εexcited to a higher energy state, they must lose their excess energy in order to return to the normal ground electronic state. Most molecules will simply heat up and lose the excess energy through heat radiation. Molecules that are rather planar and rigid and have conjugated double bonds have a greater tendency to fluoresce light. These molecules are most often referred to as fluorescent dyes. Common dyes used in molecular biology include FAM, TET, HEX, TAMRA, NED, Pacific Blue, and many others (Table 2.1).

Table 2.1. Fluorescent dyes for tagging nucleic acid fragments and the conditions for excitation and detection. Data assembled from Refs. [2, 3] and the web sites of www.SyntheticGenetics.com, www.MolecularProbes.com and www.idtdna.com.

Dyes	*Excitation Maximum (nm)*	*Emission Maximum (nm)*
6-FAM (6-carboxy-fluorescein)	494	518
JOE	520	548
TAMRA (Carboxytetramethyl rhodamine)	565	580
ROX (Rhodamine X)	585	605
Pacific Blue	416	451
Fluorescein	492	520
HEX (Hexachlorofluorescein)	535	556
TET (Tetrachlorofluorescein)	521	536
Texas Red	596	615
Cy3	550	570
Cy5	649	670
Cascade Blue	396	410
Marina Blue	362	459
Pacific Blue	416	451
Oregon Green 500	499	519
Oregon Green 514	506	526
Oregon Green 488	495	521
Oregon Green 488-X	494	517

Table 2.1. (continued)

Dyes	*Excitation Maximum (nm)*	*Emission Maximum (nm)*
Rhodamine Green	504	532
Rodol Green	496	523
Rhodamine Green-X	503	528
Rhodamine Red-X	560	580
BODIPY FL	502	510
BODIPY 530/550	534	551
BODIPY 493/503	500	509
BODIPY 558/569	559	568
BODIPY 564/570	563	569
BODIPY 576/589	575	588
BODIPY 581/591	581	591
BODIPY FL-X	504	510
BODIPY TR-X	588	616
BODIPY TMR	544	570
BODIPY R6G	528	547
BODIPY R6G-X	529	547
BODIPY 630/650-X	625	640

Fluorescence is the immediate emission of light from a molecule after it has absorbed radiation. With the absorption of the appropriate radiant energy, a molecule is raised from a vibrational level in the ground state to one of many possible vibrational levels in one of the excited electronic levels. The emission of light results from a relaxation to a lower electronic energy state from its excited higher electronic state giving a wavelength of light corresponding to the decrease in energy states. The absorption and emission of light are specific for that particular fluorescent dye molecule. A typical absorption spectrum and emission spectrum are shown in Figure 2.10. The excitation spectrum is represented by a plot of different excitation wavelengths vs. the emission signal at the optimum wavelength. The emission spectrum is measured by keeping the excitation wavelength constant and measuring the emission signal intensity at the various wavelength. The ab-

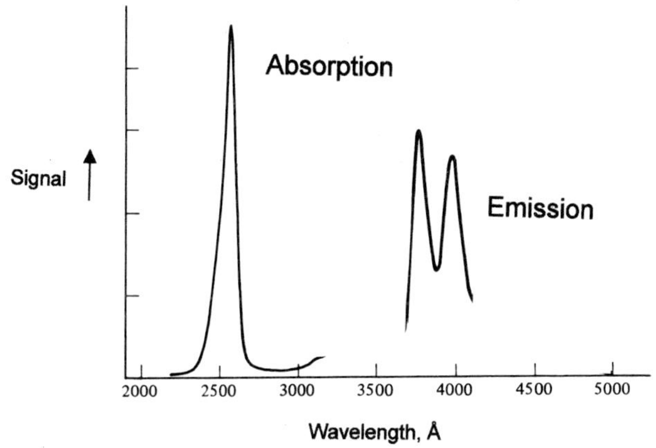

Figure 2.10. Typical absorption and emission spectra for a fluorescent labeling dye. The emission is at a lower energy and longer wavelength than the absorption.

sorption maximum is always at a shorter wavelength (higher energy) than the emission maximum (lower energy) because some energy is lost as heat.

The fluorescence process is rapid (in the order of 10^{-8} seconds after absorption) and the signal rate is high. In most cases the background fluorescence signal is low giving a high signal to noise level (a high detector sensitivity). Normally a "tagged" or "labeled" nucleic acid molecule will have only one fluorescent molecule attached to each DNA fragment. The labeling process is usually performed as part of PCR where labeled primers are used to amplify the fragment. Therefore, on a molar basis, the fluorescence is uniform for different fragments; fragments of different sizes will fluoresce the same amount because they all have one tag per fragment. On a mass basis, the amount of fluorescence per unit mass is dependent on the fragment size; a larger fragment will fluoresce less than a smaller fragment. This is different from UV detection where sensitivity is based on mass because all of the molecule absorbs light for detection. Even with only one tag per nucleic acid fragment, fluorescence detectability is 30 to 100 times greater than UV [2,3]. Stated in a different way, the detection limits are 30 to 100 times lower.

There is also another method of fluorescence tagging where fluorescent intercalating or grooving reagents are added to the sample prior to injection. In this method, all sequences of the fragment are reacted and detection is uniform. This type of tagging process will be discussed later.

Figure 2.11 shows a schematic of a fluorescence detector cell. Several configurations are possible, but for conventional detectors the excitation light source is positioned 90 degrees from the measurement. Stray light is harmful and should be avoided because this will contribute to the background signal. The detector should have variable excitation and variable emission wavelength capability. A diffraction grating is usually used to vary the wavelength. If the type of dye is not varied much, then optical filters may be used to control the wavelength. Table 2.1

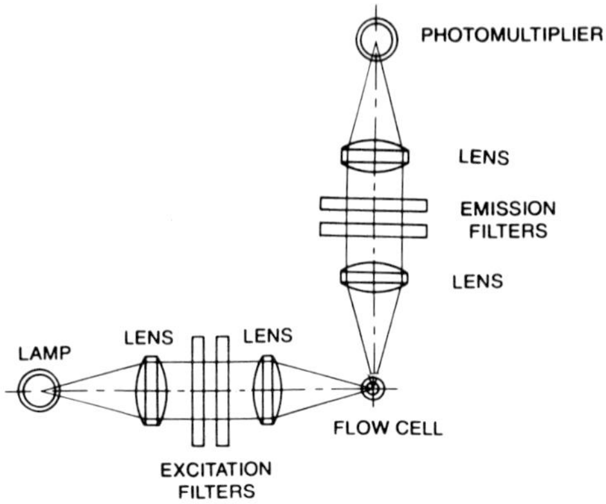

Figure 2.11. Schematic of cell for fluorescence detector.

shows the excitation and emission wavelengths of the various dyes. Four color sequencing probes are used extensively in molecular biology. Four different dyes are used to look at different aspects of the same sample. The sample does not have to be separated; the detection process for all four dyes is performed on the same lane or capillary. In the same way, four color detection are possible for DNA Chromatography. Oefner and coworkers [4] modified a laser induced fluorescence detector that was originally designed for capillary arrays. Four color detection is recorded for each capillary for up to 25 capillaries in parallel. By being able to selectively detect a particular dye in the presence of three other dyes, four samples can be run on each capillary. All dyes were excited at 488 nm. The four detection channels were set to 525, 555, 580 and 595 nm.

Theoretically, four samples can be run simultaneously on the same column. Some precautions must be taken. The fluorescent molecule will increase the hydrophobicity of fragment and this means that greater concentrations of acetonitrile are needed to elute the fragment from the column. Different labels contribute differently to the total hydrophobicity of the molecule. Since by definition four samples injected together will use the same gradient, matching the dyes is critical so that labeled fragments will elute at the same time. Three dyes that behave similarly are 6-FAM, HEX, and NED. Of course dyes will never be exactly alike, so a slightly longer gradient program may be needed to compensate for differences in retention.

2.3.11.4 Mass Spectrometry Detection

Mass spectrometers are powerful analytical tools that are able to extract a wealth of information concerning the structure of organic compounds and complex organic mixtures. The mass spectrometer produces charged particles consisting of the parent ion and smaller ionic fragments that are then sorted and recorded according to the mass/charge ratio of each particle. Each compound has a characteristic mass spectrum that consists of the different kinds and relative number of each ion.

The mass spectrometric parameters and conditions will affect the type of mass spectrum that is produced for a particular compound. A detailed interpretation of the mass spectrum record of a sample makes it possible to determine the molecular weight of the compound, and perhaps even place functional groups into certain areas of the molecule to see how the molecule is constructed. It is possible to use the mass spectrometer directly on samples without prior separation by chromatography. This works well with samples containing a single or limited number of compounds of interest. If the sample contains a number of compounds or if the sample matrix is complex, there are significant advantages to connecting the mass spectrometer to a HPLC.

A mass spectrometer consists of four parts, (1) the inlet system; (2) the ion source; (3) the analyzer system; and (4) the detector and readout system. In addition, the instrument maintains a high vacuum throughout the spectrometer from the inlet to the detector. Molecular flow of the sample fragments throughout the mass spectrometer and to the detector can only be achieved through the (almost complete) removal of gases (air) and solvents. Nucleic acids have extremely high molecular weights and are nonvolatile. Connection of a mass spectrometer to a liquid chromatograph can be a powerful combination of analytical tools. In the detection process, liquid chromatography introduces solvent and other material to the mass spectrometer. This places constraints on the mass spectrometer inlet, ionization and analyzer methods that can be used for DNA Chromatography.

The HPLC produces a continuous flow of fluid and sample that must be introduced into the sample inlet of the mass spectrometer. Consequently, it is necessary to drop the pressure experienced by the sample quickly. Since there is a limitation of how much liquid that can be actually introduced into a mass spectrometer and still maintain a suitable vacuum, it is common to use a splitter to use only a portion of the HPLC effluent. Another method uses microbore columns that require much lower flow rates for operation. For example a 1 mm bore column requires only 5 % of the flow rate that is required by a 4.6 mm bore column. In one type of inlet system, a series of inlets are arranged where the vaporized liquid is drawn into a chamber by an auxiliary mechanical pump. The low pressure in the sample inlet reservoir draws in the liquid from the HPLC and vaporizes it instantly. A heated inlet system will extend the usefulness of mass spectrometry to polar materials and high boiling compounds that are prone to be adsorbed on the walls of the chambers. The temperature that can be used is limited by the thermal degradations properties of the compound. Above 200 °C, most compounds containing oxygen or nitrogen will thermally decompose.

However, past attempts at applying mass spectrometry to DNA research have been difficult because of the problems associated with vaporizing and ionizing molecules as large as a typical DNA fragment. The ionization efficiency of a mass spectrometry source must be high so that a large portion of the neutral sample particles that are present in the inlet will become ions. There are a number of ionization methods used in mass spectrometry including electrospray, matrix assisted laser desorption, electron impact, chemical ionization, fast atom bombardment, chemical ionization, spark source, and others. Electrospray ioniation (ESI) is a gen-

Table 2.2. Types of Mass Spectrometry ionization methods.

Ionization Method	*Typical Analytes*	*Sample Introduction*	*Mass Range*	*Description*
Electrospray (ESI)	nucleic acids, proteins nonvolatile	HPLC effluent, liquid	to 200 K Daltons	soft ionization method
Matrix Assisted Laser Desorption (MALDI)	nucleic acids, proteins nonvolatile	sample mixed in solid matrix	to 500 K Daltons	soft ionization method very high mass
Fast Atom Bombardment (FAB)	carbohydrates, peptides nonvolatile	sample mixed in viscous matrix	to 6 K Daltons	medium soft ionization method
Electron Impact (EI)	small molecules, volatile	gas or liquid	to 1 K Daltons	hard ionization method
Chemical Ionization (CI)	small molecules, volatile	gas or liquid	to 1 K Daltons	soft ionization method

tle or soft ionization method (Table 2.2). A charged particle beam sprays the droplets as the mix is being evaporated. The fragments remain mostly intact as the ions are forms. The fragment ions may be positive or negative depending on the mode being used and usually has between one and three charges. Nucleic acids are naturally negatively charged due to their phosphate groups. The phosphate groups are paired with positive counter ions that are present in the chromatographic effluent so that the charge associated with the fragment is only the charge introduced by the electrospray method.

There are many different types of mass analyzers depending on the mechanism that is used to differentiate the charged ions (Table 2.3). The description and differences of the many different kinds of mass spectrometers are beyond the scope of the book. Willard et al. have written an excellent section on the subject in their analytical chemistry book [5]. The most common mass analyzer that is used for nucleic acid chromatography is the quadrupole. The quadrupole field is formed by four electrically conducting, parallel rods that are oriented symmetrically around an ion path. Equal but opposite potentials are applied to two pairs of rods, each potential having dc and rf components. An ion injected down the axis will travel the length of the rod without striking any of them only if it has a charge to mass ratio that corresponds to frequency that is applied. Changing the electrical parameters will change the mass that can be detected by the ion detector located at the end of the rods. The electrical parameters can be scanned so that as many as 1000 atomic mass units can be scanned every second. It is possible to modify

Table 2.3. A limited of some types of mass analyzers.

Analyzer	Description
Quadrupole	Unit mass resolution, fast scan, low cost, Multiple quardrpoles in series possible
Time of Flight (TOF)	high throughput
Sector (magnetic and/or electrostatic)	High resolution, exact mass
Ion Cyclotron Resonance (ICR)	Very high resolution, exact mass

this type of mass spectrometer so that a series of quadrupoles can be used to increase mass resolution.

There have been a number of papers written on separating small single-stranded DNA and detecting them with mass spectrometry (MS). Bleicher and Bayer [6] used ESI-MS to detect up to 24-mer oligonucleotides separated on a 100 mm long $\times$ 2 mm inside diameter Nucleosil^{TM} C-18 column with gradient of acetonitrile solvent with 10 mM ammonium acetate ion pairing reagent. Bothner et al. separated oligonucleotides using an eluent containing 200 mM diisopropylammonium acetate in acetonitrile on a 150 mm long x 5 mm inside diameter column packed with a polymer PLRP-S stationary phase [7]. Phosphodiester, methylphosphate, and phosphorothioate oligonucleotides up to 20-mer in length were characterized at the 7 nmole level. Apffel et al. utilized a 250 mm long by 2.1 mm inside diameter YMC C-18 column to analyze 75-mer oligonucleotides [8,9]. They found that 100 mM triethylammonium acetate ion pairing reagent resulted in a drastic reduction of ion formation giving poor mass sensitivity. Performance was improved by using a sheath of organic solvent around the eluent fluid to improve the stability of the electrospray and to enhance signal intensity of the ESI-MS.

The detection of nucleic acids by electrospray ionization can be difficult due to their tendency to form stable adducts with cations such as sodium resulting in mass spectra of poor quality. Removal of sodium cation is necessary in order to obtain high quality mass spectra. The counter ion of the ion pairing reagent can also drastically affect the result. For example, a multivalent counter ion will cause the fragment to elute faster from the column. Some of the best work has been performed by Huber and coworkers where conditions for separation and detection have been optimized [10, 11].

2.3.12 Data Analysis

The results of the chromatographic separation are generally displayed on a computer screen. The computer uses an A/D (analog to digital) board to convert the analog signal coming from the detector into digital data. The digital information is stored and manipulated to report results to the user. There are three types of information that can be gained by the peaks of the chromatogram. They are the peak retention time, the peak shape or pattern, and the peak size (area and height).

2.3.12.1 Size Analysis

DNA fragment sizing is required for quality control of sequencing templates, sizing of DNA fragments from PCR or enzymatic digests, and for optimization of PCR conditions [12,13]. When working with double-stranded DNA, reverse phase columns and ion pairing reagents such as triethylammonium acetate (TEAA) will product size based separations of the fragments. The retention times from one column to the next and from one day to the next are quite reproducible provided the column is in good condition and the eluent has been prepared properly and is not too old. The working status of a system can be determined by injecting a pUC18 *Hae* III digest (available from Transgenomic, San Jose, CA and Omaha, NE) and comparing the chromatogram to the separation that comes with the standard. It is possible to perform what is called zero point calibration of the system to determine the size of an unknown fragment. WAVEMaker® system software (Transgenomic) has default retention time values in equations that are used to calculate the size of an unknown. It is possible to accurately calculate fragment size within 10% with systems that are maintained. Greater accuracy will be achieved using 1 point calibration with a single known size standard, or better yet, with multipoint size calibration (a DNA ladder or enzymatic digest). Some commercial DNA ladders are not suitable for DNA Chromatography because of impurities, thus it is recommended to use only standards from the manufacturer.

Separations of single-stranded DNA and RNA do not proceed only according to fragment size. The fragment sequence also contributes to the retention time. Generally early eluting peaks will correspond to shorter (degraded) fragments. In these cases, it probably is not necessary to run a standard unless there is some reason to confirm the identity of the major peak.

2.3.12.2 Peak Shape or Pattern of Peaks

In many cases, it is only necessary to observe difference in peak shape or peak pattern of the sample versus a standard. DHPLC is the separation of heteroduplex and homoduplex species. A sample mixture of heteroduplices and homoduplices normally contains four species and the ideal separation of this sample is a four peak pattern. However, it is uncommon to get this type of pattern. More likely the pattern is 3 or 2 peaks or even an unresolved peak in front of the homoduplex peak (see Chapter 4). The most efficient method for analyzing data of this type is to normalize the last peak and overlay all the peaks so that the back part of the last peak line up. Any differences that indicate the presence of a heteroduplex are seen as either a broadening of the front part of the peak or extra peaks in the front part of the chromatogram (see Figure 2.12).

There is an interesting side note to this. Many researchers performing mutation or SNP (single nucleotide polymorphism) analysis do not even use standards. The

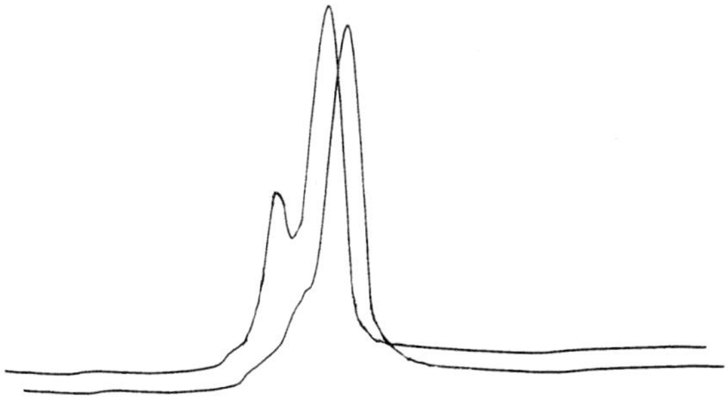

Figure 2.12. Overlay of a wild type and a mutation chromatogram. DHPLC chromatograms normally have a rising baseline before the main last peak due to PCR induced mutations. This can make identification of some low resolved mutations difficult. But, by normalizing the last peak and adjusting retention times of the back side of the last peak, even small differences between the wild type and mutation chromatograms can be observed.

wild type samples (the homozygous standard) are chosen after the analysis has been performed. Samples with the narrowest and most consistent peaks are the wide type samples. Heterozygous samples will produce a heteroduplex sample and can be detected with the characteristic multiple peaks. Of course, analysis of this type will miss homozygous mutations, but this might be acceptable for mutation discovery work.

Peak patterns may also be important in DNA or RNA footprinting applications (discussed in Chapters 5 and 7). In these cases, it is differences in the shape and number of peaks over the whole chromatogram that are measured and examined.

2.3.12.3 Quantification

There are many cases where it desired to know how much of a particular fragment of DNA or RNA is present. Downstream processing such as PCR, cloning, reverse transcription, etc. are done much more efficiently if the amount of nucleic acid is known. The absorptivity for double-stranded DNA is about 80 % of single-stranded DNA or RNA so that for a given absorbance, the sample will contain more DNA than RNA.

For a given instrument configuration, it is possible that a statement could be made about the area/mole ratio for a typical analyte. Much better results could be obtained if a standard was run at least once on the configuration to be used to determine the area/mole relationship. If the instrument is working properly it is possible to calculate concentration at least to within 50 % with no (zero point) instrument calibration. More accurate work requires the use of a standard of known concentration.

Recall that the whole molecule is detected with UV detection and either molar or mass concentration or mass amount can be calculated. With fluorescent detection usually molar concentration is calculated. (The mass concentration can be calculated using the molecular weight of the particular fragment being measured). Using a simple ratio the calculation of the concentration can be performed:

$$\text{unkn. conc. / unkn. peak area} = \text{kn. conc. / kn. peak area} \quad (2.2)$$

therefore,

$$\text{unkn. conc.} = \text{kn. conc. / kn. peak area x unkn. peak area} \quad (2.3)$$

It is also possible to draw a calibration curve of known peak area vs. known concentration, finding the unknown peak area on the curve and measuring the unknown concentration. Much of this is performed automatically by software once a method is installed. The multipoint calibration curve is the most accurate method, but rarely is this level of accuracy needed and single point calibration ratio is usually all that is used.

In order for the data system to measure peak area or peak height, the baseline of the peak must be accurately drawn. The software program will attempt on its own to draw a baseline for the peak, but frequently the user must manually mark the baseline start and finish points to accurately draw the peak baseline. It is a common mistake of new users to trust the software to draw the correct baseline. Figure 2.12 shows the correct way to draw baselines for peak integration (measurement of peak area).

It is often necessary to convert moles of material concentration to mass of material. For double-stranded DNA conversion of pmol to µg:

$$\text{pmol} \times \text{N} \times 660 \text{ pg/pmol base pairs / 1x } 10^{-6} \text{ pg/µg} = \text{µg} \quad (2.4)$$

where N = number of base pairs.

Sample preparation and injection volume errors probably account for most of the short term variation in area measurement. Poor pipette techniques can also create some error.

2.3.13 Fragment Collection

One of the powerful features of DNA Chromatography is that material can easily be purified by collecting directly from the detector effluent. The collection can be done by hand, but is normally accomplished with an automated fragment collector and controlling software. Collection occurs into single vials or large PCR plates.

Care should be taken to ensure that the actual peak has been collected. Measurement of recovery is performed by taking a small portion of the recovered peak, re-injected and measuring the area, multiplying this by the ratio of total collected volume to reinjected volume and comparing this value to the area of the original peak. Normal recoveries are about 80 %. Of course materials such as RNA may degrade (by enzymes) after they have been collected. High recoveries may not be possible due to loss of the material during the concentration process (precipitation or solvent evaporation).

An important parameter in fragment collection is careful execution of the collection times. There is a lag from the time the peak is seen in the detector cell and when the fragment is deposited from the fragment collector probe. The most reliable collection method is the timed collection. But unless this is timed correctly, it is easily possible to miss some or all of the peak. It is also important that there is as little dead volume in the tubing from the detector cell outlet to the tip of the deposition probe. Too large of a dead volume will destroy the resolution of the separation and could result in cross contamination of the peak of interest with neighboring peaks [14]. It is important that the probe is cleaned between collection of peaks. This is normally done automatically by the fragment collector.

References

1. J. E. Sutton, D. T. Gjerde, and P. D. Taylor, MIPC chromatographic apparatus with improved temperature control, U. S. Pat., 6,103,112, 2000.
2. P. J. Oefner, C G. Huber, F. Umlauft, G. N. Berti, E. Stimpfl, and G. K. Bonn, High-resolution liquid chromatography of fluorescent dye-labeled nucleic acids, Anal. Biochem., 223, 39, 1994.
3. K. H. Becker, P. D. Taylor, and D. T. Gjerde, Mutation detection by denaturing DNA chromatography using fluorescently labeled polymerase chain reaction products, Anal. Biochem., 272, 156, 1999.
4. W. Xiao, D. Stern, M. Jain, C. G. Huber, and P. J. Oefner, Multiplex capillary denaturing high-performance liquid chromatography with laser-induced fluorescence detection, BioTechniques, 30, 13332, 2001.
5. H. Willard, L. Merritt, J. Dean, and F. Settle, Instrumental methods of analysis, 7th ed. Wadsworth Pub.: Stamford, CT, 1998.
6. K. Bleicher and E. Bayer, Analysis of oligonucleotides using coupled high performance liquid chromatography – electrospray mass sepectrometry, Chromatographia 39, 405, 1994.
7. B. Bothner, K. Chatman, M. Sarkisian, and G. Siuzdak, Liquid-chromatography mass-spectrometry of antisense oligonucleotides, Bioorg. Med. Chem. Lett., 5, 2863, 1995.
8. A. Apffel, J. A. Chakel, S. Fischer, K. Lichtenwalter, W. S. Hancock, Analysis of oligonucleotides by HPLC – electrospray-ionization mass spectrometry, Anal.Chem., 69, 1320, 1997.
9. A. Apffel, J. A. Chakel. S. Fischer, K. Lichtenwalter, W. S.Hancock, New procedure for the use of high-performance liquid chromatography – electrospray mass spectrometry for the analysis of nucleotides and oligonucletides, J. Chromatogr., 777, 3, 1997.
10. C. G. Huber and A. Krajete, Analysis of nucleic acids by capillary ion-pair reversed-phase HPLC coupled to negative-ion electrospray ionization mass spectrometry., Anal. Chem., 17, 3730. 1999.
11. A. Premstaller, H. Oberacher, and C. G. Huber, High-performance liquid chromatography-electrospray ionization mass spectrometry of single and double-stranded nucleic acids using monolithic capillary columns, Anal. Chem., 72, 4386, 2000.
12. C. G. Huber, P. J. Oefner, and G. K. Bonn, Rapid and accurate sizing of DNA fragments by ion-pair chromatography on alkylated nonporous poly(styrene-divinylbenzene) particles, Anal. Chem., 67, 578, 1995.
13. K. Munson, R. Haefele, A. Kuklin, D. Gjerde, and P. Taylor, Sizing of DNA fragments with the WAVE® Nucleic Acid Fragment Analysis System, Application Note 107, Transgenomic, Omaha, NE.
14. K. H.Hecker and K. Kobayashi, Sequencing of DNA fragments iolated with the WAVE® Nuceic acid fragment analysis system, Application note 115, Transgenomic: Omaha, NE.

3
Chromatographic Principles for DNA Separation

3.1
Introduction

The Russian botanist Mikhail Tswett coined the term "chromatography" in 1903 from the Greek words meaning "color" and "writing" [1]. He described a method for the separation of pigments found in plant leaves using a tube or column filled with a dry solid adsorbent of granular calcium carbonate (chalk) and other absorbents. He added an extract of plant material containing its pigments to the top of the column and then washed the column with an organic solvent. As he washed the mixture of plant pigments down the column, the pigments separated into a series of discreet colored bands on the column between regions entirely free of pigments. After all of the bands were resolved (separated), he pushed the moist chalk material out of the tube as an intact cylinder and recovered each pigment by cutting the bands apart with a knife.

Tswett's development of chromatography procedures was generally unknown to chemists in the Western world because he published in German botanical and Russian journals. But he was the first to recognize the physiochemical basis of chromatography and apply it in a rational and organized way in the separation of plant pigments. He recognized that chromatography is based on the relative attractiveness of compounds in solution to either "attach" to a stationary (solid) phase of the column or to be dissolved by a mobile (liquid) phase.

This attachment is only a partial attachment. Most compounds are attracted to both the mobile phase and stationary phase. Separation of materials by chromatography occurs when the components of a mixture differ in their attraction to the two phases. Materials in the mixture that don't "like" the mobile phase but rather "like" the stationary phase stick more tightly to the column material and elute or travel slowly down the column. Materials that "like" the mobile phase elute or travel down the column more quickly. This process continues until all of the components of the mixture are separated. (Figure 3.1).

After Tswett's original work, it was some time before the significance of his work was realized. But eventually more research was done [2–4], and it was discovered that chromatography can be used for many different types of applications including separation of proteins, organic and inorganic ions, small organic molecules,

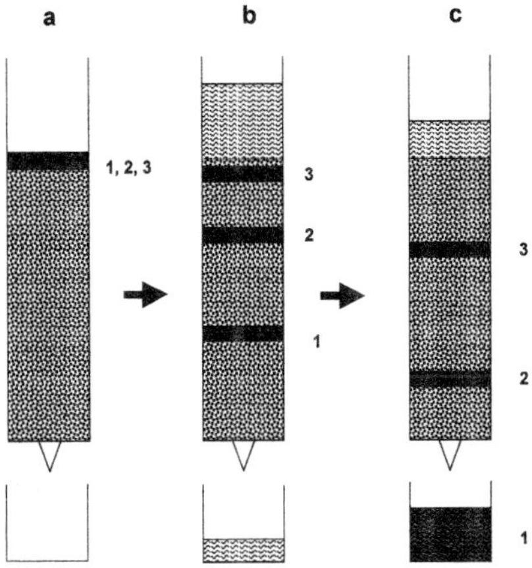

Figure 3.1. The basic separation process. a) Components of a mixture are added as a concentrated aliquot to the top of the column. b) Adding eluent to the top of the column starts the separation process. Some components will travel through the column faster because they are solvated by the eluent and interact less with the column packing. Other components interact more strongly with the packing and travel slower through the column. c) Component 1 is eluted and collected.

carbohydrates, amino acids, and of course DNA. Depending on the application, the column is packed with a suitable material usually from a group of materials that are either silica-based or polymer-based. Many different types of eluents are used including salts, buffers, acids, bases, and different organic solvents. Each type of separation is based on a particular type of liquid mobile phase, which creates the correct interaction between the two phases.

Chromatography also expanded into analogous technologies using other types of stationary and mobile phases. The most successful and popular is gas chromatography, where the column stationary phase material is solid and a gas mobile phase such as helium, or nitrogen is passed through the column. Volatile substances such as alcohols are injected into the gaseous mobile phase and the alcohols are separated as they move along the column. Temperature is used in these separations as an added parameter. During the chromatographic run, the temperature is gradually increased, helping the alcohols to increase their vapor pressure and associate with the mobile phase as they pass through the column. The whole process is related to distillation where the most volatile alcohols will elute first off the column and then the higher boiling alcohols elute from the column. We shall see that using the analogy to distillation will become useful when describing the chromatographic process.

Classical liquid chromatography is still practiced today and carried out in much the same way as Tswett's original work. Glass columns are filled with a suitable solid stationary phase and then eluent mobile phase is added to the top to condition the column material. The liquid flows through the column and drips out the bottom with the flow rate usually controlled with a stopcock. The mobile phase cleans and conditions the column. The sample is added to the top of the column in a single, concentrated slug and then the washing solvent is again added to the top of the column washing the sample components down the column and into collection vials. If the sample components are colored, visual record can be used to track the separation and collect the materials. In other cases, fractions of effluent are collected from the end of the column and analyzed. A plot is made of concentration vs. fraction number to construct a chromatogram.

It was realized that under a given set of conditions, a particular sample component could always be found in a certain volume fraction of the eluent, the next compound could be found in the next fraction and so on. For example, Compound A is always located in 3–5 mL eluent fraction, Compound B is located in the 5–7 mL fraction and Compound C is located in the 7–11 mL fraction. Thus, under pre-determined conditions, even non-colored components of mixtures can be separated and collected in a single fraction and then analyzed by spectroscopy, titration, etc. to determine the amount of each component in the original mixture. Still, all this takes a long time and makes column liquid chromatography slow and awkward to use.

Starting in the late 1960s, liquid chromatography changed suddenly and drastically with the development of improved theories and instrumentation. The first major scientific meeting to discuss the subject, The 5th International Symposium on Advances in Chromatography, was held in Las Vegas on January, 20–23, 1969. Previous meetings of this symposium discussed the technique of gas chromatography, but this was the first major meeting where papers (sixteen total) were presented on modern liquid chromatography. However, the name "high-performance liquid chromatography" did not exist at that time as yet. C. Horvath and S. R. Lipsky had a joint paper where the technique was called "high pressure liquid chromatography." The term "high-performance liquid chromatography" with the acronym HPLC was coined by Csaba Horvath in a paper presented at the 21st Pittsburgh Conference held in Cleveland, OH, March 1970. This was the title of his lecture and the name HPLC was adopted almost immediately by everybody [5, 6].

HPLC did not really become fully commercialized until the late 1960s and early 1970s when improved columns of silica packing materials with bound layer of organic alkyl groups were introduced along with improved automated pumps and detectors. These new packing materials were small (10 μm average particle size) and packed uniformly into small stainless steel columns. The column separated materials in very tight bands allowing more bands to be separated and smaller elution volumes to be used. Gradually gradient elution technology was used to improve the separations. Gradient elution uses a mixed solvent eluent system that changes concentration over time producing a gradient. The eluting power of the mobile phase increases as the separation proceeds thus speeding the elution of materials that would normally be held very tightly by the column.

As described in the last chapter, liquid chromatographic instrumentation systems use a pump, injector, column and detector. The pump makes it possible to automatically pump the eluent fluid through the column precisely (uniformly without minimal pulsations), accurately, and under high back pressure conditions (normally one or two thousand psi). The injector makes it possible to perform precise injections (the same amount and time) into the fluid stream entering the column. Finally, the detector allows continuous detection and automatic recording of the separation.

HPLC has become more and more popular because the new technology has provided reproducible and quicker separations. While instrumentation holds a large part to the success, the key to making HPLC work is the column and the interactions of the components of the mixture with the column and the eluent. Much of the largest breakthroughs in chromatographic research have been with the column. In HPLC, smaller and more efficient columns are used. Classical gravity columns are 300 to 500 mm long and have inside diameters between 8 and 25 mm. Newer HPLC columns are only 150 to 300 mm long and can be as short as 30 mm. The most common inside diameter of an HPLC column is 4.6 mm but diameters can be as small as 1 mm. The particle size of packing material in the columns is also much smaller. Classical materials are 25 to 100 micron diameter whereas most HPLC columns are 5 micron diameter. The trend to even smaller particles and column dimensions continues today where DNA separation column packings are typically 2 micron.

This chapter discusses the principles of chromatography. While instrumentation is important, the key to the separation process is the chemistry that occurs with the column (the stationary phase) and eluent (the mobile phase). The column and eluent performance requirements of the components systems used in modern liquid chromatography are summarized. Modern HPLC employs:

1. An efficient column with as high of a resolving power as possible.
2. An eluting solvent or eluent that provides reasonable separation of the mixture.
3. A column/eluent system that attains equilibrium quickly so that peak broadening due to slow chromatography kinetic interaction rates is eliminated or minimized.
4. Elution conditions such that retention times are in a convenient range – not too short or too long.

Chromatography can be explained by a description of multiple chemical processes. The processes can be expressed through mathematical equations. Much of the mathematical information in this chapter applies to HPLC in general, while other chemical interaction information applies only to DNA Chromatography. Although the reader does not need to know or understand all information in this chapter to use DNA chromatography, better understanding will help the reader to use the technology efficiently or even find new uses for DNA Chromatography

3.2
Comparison of Chromatography and Gel Electrophoresis

Gel electrophoresis is the fundamental analytical technology upon which modern molecular biology has been built. It is a staple tool in molecular biology and critical in the success of many aspects of genetic manipulation and study.

The key aspect of gel electrophoresis is the gel material. When a potential difference voltage is applied across the electrodes, voltage potential gradient is generated through the gel. It is this force that drives or attracts the charged nucleic acid molecules to the electrode. The negative charges of the nucleic acid phosphate groups are attracted to the positive electrode (anode) of the apparatus.

There is also a frictional resistance that slows down the movement and provides the separation selectivity of the charged nucleic acids. This friction force is a measure of the hydrodynamic size of the molecules, the shape of the molecule, and the pore size of the medium (gel) in which the electrophoresis is taking place. Since the hydrodynamic size of the molecule is closely related to the number of bases or base pairs, the separations are sized based.

The gel consists of long polymer chains that are hydrated or swollen with water with more than 95 % of its weight in water. There are two basic types of material used to make gels: agarose and polyacrylamide. Purified agarose is in powdered form and is insoluble in water or buffer at room temperature but will dissolve in boiling water. When it starts to cool, it undergoes polymerization and thickens. The resulting polysaccharide has a molecular weight of approximately 12,000 and is used at concentration between 1 and 3 %. Agarose gel electrophoresis is faster than polyacrylamide gel electrophoresis (PAGE), but the resolution obtained is poorer. Agarose is a natural colloid extracted from sea weed. It is very fragile and easily damaged by handling. Agarose gels have very large effective pores and are used primarily to separate very large nucleic acid molecules up to 20,000 base pairs.

Polyacrylamide gels may be prepared as to provide a wide variety of separation conditions. The pore size for the gel may be varied to produce different molecular sieving effect for separating DNA of 50 to 10,000 base pairs. Polyacrylamide gels can be cast as a single percentage or with varying gradients. Gradient gels provide continuous decrease in pore size from the top to the bottom of the gel, resulting in thin bands. Polyacrylamide gels offer greater flexibility and more sharply defined banding than agarose gels, but are much more difficult to prepare and handle.

It has been known for years that ions can be separated by differences in their rates of movement in an applied electric field (electrophoresis). Until fairly recently, electrophoresis was considered to be rather slow. The realization that electrophoresis can be performed in a fused silica capillary has resulted in some dramatic changes. Now DNA and RNA can be separated at high resolution in an automated method. However, separation times are still normally long and only small amounts of materials can be separated and purified.

Despite the ability of DNA Chromatography to perform size-based separations of DNA, it is an entirely different process compared to gel electrophoresis. The chromatographic separation takes place by pumping a fluid (eluent) through a column. DNA is injected into the fluid stream and DNA sticks to the column. Then the eluent fluid strength is gradually increased until the DNA fragments are eluted from the column, starting with the smallest fragments and ending with the largest. After the separation is completed, the eluent is returned to the original strength and the whole process can be repeated.

Details of the column and how is used are described in the next sections, but it can be seen that there are many differences of DNA Chromatography to gel electrophoresis. The major difference is the column can be thought of as a "programmable gel". The column is programmable in several different ways. One, the same column can perform any number of separations simply by choosing the appropriate set of elution conditions. Large fragments 500–1000 base pairs can be analyzed in one run, and 5 minutes later in the next run, a 100 base pair fragment can be measured. Also a mixture of 100–1000 base pair DNA can be analyzed in still another run. The column performs a separation as a response to the elution conditions that the column is undergoing at that instant. Thus, a gradient of elution conditions can produce any desired response. Once a gel is made and the process begins for gel electrophoresis, no changes in the separation can be made.

DNA Chromatography is fast. An experiment can be conceived, prepared and analyzed within a matter of minutes – making the technology perfect for "what if" type R&D experiments. The major benefits of gel electrophoresis are that it can be multiplexed (more than one channel can be run simultaneously), it can be used for sequencing, and large fragments can be analyzed.

Table 3.1. Comparison of DNA Chromatography and Gel Electrophoresis

	DNA Chromatography	*Gel Electrophoresis*
Separation Force	solubility of analyte in eluent flowing through column	electromotive attraction of analyte to electrode
Separation Resistance	attraction of analyte to column stationary phase	resistance of analyte to travel through gel pores
Sample Separation Property	hydrophobicity or ion exchange	hydrodynamic diameter (volume)
Primary Separation Effect	size – can also be controlled to include sequence effect	size
Size Range	small to medium	medium to large
Detector Sensitivity to Mass Amount	excellent	needs fluorescent tag
Effect of Separation Temperature	partial and full denaturing	difficult to use
Gradient to Control Separation Selectivity	yes	difficult
Automation	yes	yes with capillary – purification still difficult
Separation Speed	fast	medium to slow
Multichannel	no	yes
Purification	yes	difficult

DNA Chromatography is not a replacement for gel electrophoresis. The two techniques are functionally similar because the separations are both size-based. But as we shall see, the mechanisms for separation are quite different for the two techniques. It is these differences that can translate into advantages for DNA Chromatography for many applications. Both technologies have their place in the modern molecular biologist's laboratory. A comparison of the technologies is shown in Table 3.1. Many of the applications discussed in this book take advantage of the unique aspects of DNA Chromatography. The reader is invited to examine their own research and determine which technology best suits their own needs.

3.3 Basic Chromatographic Considerations

The major chromatography terms that describe performance are summarized in Figure 3.2 and Table 3.2. Derivation and use of these terms are described in the next several sections.

Table 3.2. Chromatographic Terms

Term	*Symbol*	*Definition*
Retention time	t or t_R	Time to elute a peak to its maximum height
Dead time	t_O or t_M	Time to elute a non-sorbed marker
Adjusted retention time	t'	$t' = t - t_O$
Retention factor (or capacity factor)	k or k'	$k = \frac{t - t_o}{t_o}$
Peak width	w	Peak width at its base usually in time units
Peak width at 1/2 peak height	$w_{1/2}$	Peak width at half height usually in time units
Peak resolution	R_S	$R_s = \frac{2\,\Delta t}{\text{peak 1 w + peak 2 w}}$
Separation factor	α	$\alpha = \frac{k_2}{k_1} = \frac{t'_2}{t'_1}$
Theoretical plate number	N	$N = 16\left(\frac{t}{w}\right)^2 = 5.54\left(\frac{t}{w_{1/2}}\right)^2$
Height equivalent of a theoretical plate	H	$H = \frac{L}{N}$ (L = column length)

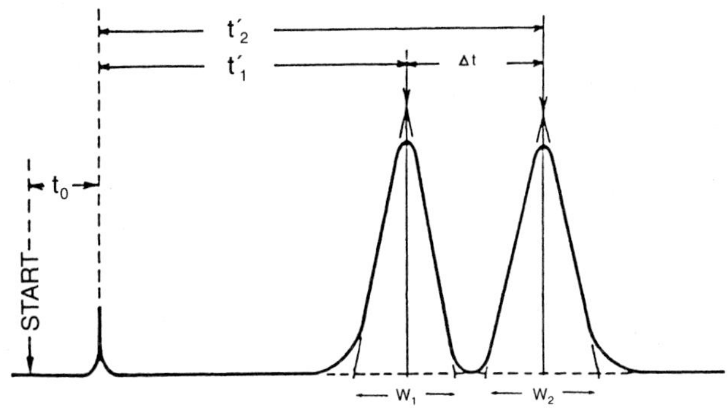

Figure 3.2. Two chromatographic peaks and their relationship to some chromatographic terms. The retention time, t, is the sum of the void time, t_o, and the adjusted retention time, t'. See Table 3.1 for a listing of terms and definitions.

3.3.1
Retention

During their passage through the column, sample molecules spend part of the time in the stationary phase and the other part in the mobile phase. All molecules spend the same total amount of time in the mobile phase. This time is called the column dead time or holdup time t_o. This time is measured by injecting a completely unretained compound and measuring the time required for the peak measured at peak height to reach the detector. The retention time, t, is the time starting at the instant of sample compound is introduced to when the detector senses the maximum of the retained peak. This value is greater than the column holdup time by the amount of time the compound has spent in the in stationary phase and is called the adjusted retention time (t') These values lead to the fundament relationship describing retention time.

$$t = t' + t_o \tag{3.1}$$

or as it is usually expressed:

$$t' = t - t_o \tag{3.2}$$

Some refer to this quantity as the dead time, t_o, while others call it the time of mobile phase to pass, t_M. Although the retention time (the time for a sample component to be eluted from the column to its peak maximum) has been given the symbol t_R, we prefer to simply write it as $\underline{t}$. This makes it possible to denote the retention times of several peaks as t_1, t_2, t_3, ...

Retention is usually measured in units of time but may also be measured in volume. The volume of eluent required to elute a substance from a column to its peak maximum is called the retention volume V. With the conversion equation, volume can be substituted for time for any equation:

$$V = t\,F \qquad (mL) = (min)\left(\frac{mL}{min}\right) \tag{3.3}$$

where F is the volumetric flow rate and

$$V = V' + V_o \tag{3.4}$$

or

$$V' = V - V_o \tag{3.5}$$

3.3.2
Retention Factors

Perhaps the single most important term in liquid chromatography is the retention factor (or capacity factor), k.

$$k = \frac{t - t_o}{t_o} \tag{3.6}$$

or

$$k = \frac{V - V_o}{V_o} \tag{3.7}$$

It is recommended to use the name *retention factor* for what was called the *capacity factor* for many years. Both *k* and *k'* have been used as the symbol for this term.

Conditions must be adjusted so that there is a sufficient difference in the k values of the various compounds in the mixture to give a good separation from each other. It is also necessary to select conditions so that the range of k values for each individual compound is such that a separation may be completed within a reasonable time. A k range of 2 to 10 has often been specified as desirable.

3.3.3
Peak Width

Figure 3.3 shows the various measurements of peak width. The chromatographic peak is assumed to be symmetrical and Gaussian shaped. Since this is not true with real peaks due to some peak tailing, it is easier to measure the peak width at 10 % height or at 1/2 peak height. Retention and peak width are used to calculate various performance measurements. In this way, the separation power of a system can be calculated and compared. However, in order to do this, the chromatographer must rely of two fundamental theories of chromatography. They are the plate theory and the rate theory of chromatography. The theories presented in this book are quite basic and have been simplified for the purpose of explaining fundamental procedures. More detailed and expanded theoretical approaches can be found elsewhere [2–4].

Figure 3.3. The peak width can be measured in different ways. Most preferred is measurement of the peak width at half height. An estimation of peak width can be made: $w = 1.7 \times w_{1/2}$.

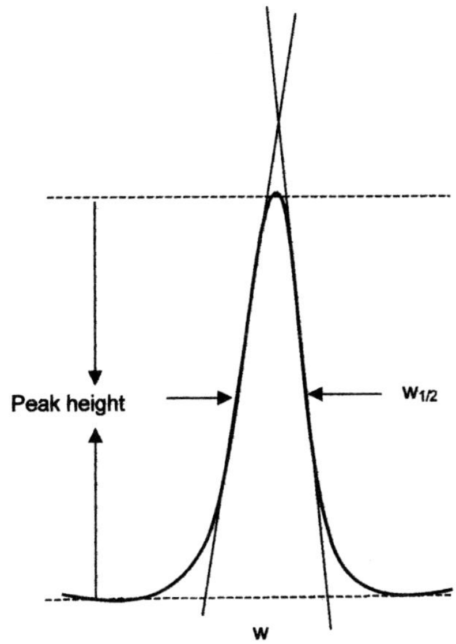

3.3.4
Plate Theory of Chromatography

In effect, it can be said that a chromatographic column provides for a large number of simultaneous equilibrations of sample material between the mobile and stationary phases. The plate theory of chromatography first attempts to define the chromatographic process through equations that describe a single equilibration of a single sample between the stationary and mobile phases. Later, the equations are expanded to assume multiple equilibrations in a column separations. Each mathematical equilibrium is said to be a theoretical plate.

Much of the first chromatographic work was done with column packing materials where an insoluble stationary liquid phase was coated onto the solid packing material. In this case, it can be said that the sample material partitions between two liquid phases: one stationary and the other mobile. For this reason we shall call the sample compound a solute. The retention volume of solute A eluting from a column can be calculated. A concentration distribution coefficient D_C is defined as:

$$D_c = \frac{[A]_S}{[A]_M} = \frac{(\text{mmoles}A)_S/V_S}{(\text{mmoles}A)_M/V_M} = D_m \frac{V_S}{V_M} \tag{3.8}$$

The subscripts S and M refer to the stationary and mobile phases respectively; and $[A]_S$ and $[A]_M$ are the concentrations of the solute A in each phase. V is the volume of a liquid phase and V_S and V_M are the volumes of liquid in each phase. D_m is the mass distribution ratio (the ratio of the amount of solute in the stationary phase to the amount of solute in the mobile phase).

Assuming that solute retention is defined by the relative amount of solutes in the two phases, it can also be seen that the following equations describe the elution volume of a solute. Taking Eq. 3.4 from above V_M may be substituted for V_o, the dead volume for a non-retained fragment. (This makes the assumption that all of the dead volume is within the column. This is not exactly true because there is some dead volume within the tubing and connections; but the values are near enough for our purposes here). The increase in elution volume is directly proportional to the ratio of the solute in the stationary phase over the mobile phase multiplied by the volume of the mobile phase. Thus, D_m multiplied by V_M may be substituted for V'.

$$V = V_M D_m + V_M \tag{3.9}$$

Another way of expressing this through rearrangement is:

$$D_m = \frac{V - V_M}{V_M} \tag{3.10}$$

Equation 3.10 is quite similar to Eq. 3.7 above since V_o and V_M are for our purposes said to be the same. Combining Eq. 3.8 with Eq. 3.10 gives the following:

$$V = V_C D_S + V_M \tag{3.11}$$

D_c can be measured in a single equilibrium experiment. Column packing material is equilibrated with a known volume of solvent and a known concentration of solute. Measurements are made of the initial solute concentration and equilibrated solute in solution. Assuming all of the solute that has gone from solution is now on the column packing material, the D_c can be calculated. The reader will note that D_m can also be calculated either with single equilibrium Eq.s or by using column parameters. Once the retention volume is determined, the retention time is calculated using the flow rate.

Of course the separation of two or more compounds requires conditions where the distribution ratios are different. How different must they be to achieve complete separation? That depends on two parameters: the difference of the distribution ratios and the performance of the system. One method to achieve separation is to choose conditions where the distribution ratios are very different. Materials with a low distribution ratio will elute early in the chromatogram and materials with larger distribution ratios will elute later. This can be quite effective; however, larger distribution ratios will increase the retention times so much so that if they are too large, it may take hours for a peak to elute from the column. In cases where the distribution ratios are quite different, a gradient must be used to get compounds to elute in the reasonable (short) time. The use of gradients will be discussed later. The other method of achieving separation is to use a column system that has a high number of theoretical plates.

Recall, that the plate theory assumed only a single equilibration or one theoretical plate to predict the retention time of a compound. As the number of theoretical plates are increased in a column, the peak becomes narrower. Narrower peaks allow the separation of more compounds within one run. Fortunately, a flowing column system has many column interactions of the same type meaning that the number of theoretical plates of a column can be quite large. Extremely high, theoretical plate numbers have been reported in the literature – up to 1,000,000. But in most cases a chromatographic system has between 2000 and 15,000 theoretical plates with which to work.

So there are really two ways to accomplish a separation. One is to choose conditions so that there is a large difference in Dc (which means there is also a large difference in the retention factor k). The other is to use a system that has a large number of theoretical plates and can therefore tolerate very small differences in retention factors. Most chromatographers use N (or sometimes n) to denote the number of theoretical plates, which is used as a measure of the separation power of a particular chromatographic column and system. A larger N indicates a greater resolving or separation power of the column in that particular system. The height equivalent of a theoretical plate H (or sometimes h) denotes the separation efficiency of a column. Thus a lower value for H denotes a more efficient column.

It should be pointed out that H and N also vary somewhat with the retention factors of the analytical sample components. Analytes with low k values generally have higher plate numbers (sharper, better resolved peaks) than those with higher k values (4 to 10, for example). It is also well known that H (and therefore N) changes with the linear flow rate (μ) of the eluent or mobile phase. The relationship will be discussed with the rate theory of chromatography below, but H generally decreases with μ.

Another important variable in HPLC is "peak capacity" which determines the number of peaks that can be present in the chromatogram. Guidelines for peak capacities as a function of the number of theoretical plates can be given for some typical ranges of retention factors. A typical packed column with 5000 plates turns out to yield a peak capacity between 17 for k from 0.2 to 2 and about 50 for k from 0.5 to 20. Note that as retention factor increases, the peaks become broader reducing the peak capacity, but a larger range of retention factors will yield a system with larger peak capacity.

As we have stated, a satisfactory chromatographic separation depends on having a column with a sufficient plate number, N, as well as an adequate difference in k values. Resolution in terms of separation factor ($\alpha = k_2/k_1$), the average retention factor, $k_{av} = (k_1 + k_2)/2$, and the plate number, N, is given by:

$$R_s = \frac{\alpha - 1}{\alpha + 1} \cdot \frac{\sqrt{N}}{2} \cdot \frac{k_{av}}{1 + k_{av}} \tag{3.12}$$

This relationship is often used to estimate the number of plates needed for a separation.

$$\sqrt{N} = \frac{2(\alpha + 1)}{\alpha - 1} \cdot \frac{1 + k_{av}}{k_{av}} \cdot R_s \tag{3.13}$$

For example, if $k_2 = 4.2$ and $k_1 = 3.8$, $\alpha = 1.08$, for $R_s = 1.0$

$$\sqrt{N} = 2 \cdot \frac{2.08}{0.08} \cdot \frac{5}{4} \cdot 1.0 = 65$$

$N = 4{,}200$ plates

In reverse phase liquid chromatography, separations are based on differences in partitioning of sample compounds between a rather hydrophobic stationary phase and a mobile liquid phase such as acetonitrile/water. Retention factors generally increase as the molecular size and hydrophobicity of the organic molecules become larger. The values of k become smaller as the proportion of organic solvent in the mobile phase is increased. The retention factors of the analytes to be separated are kept in a desirable range by adjusting the composition of the mobile phase.

In ion exchange chromatography the retention factor of an analyte is again determined by its relative affinity for the stationary and mobile phases but the mechanism is different. The stationary phase is an ion exchange material, and the sample ions are retained only at specific ionic sites on the ion exchanger. In order for the sample to be attracted to the exchange site, an eluent ion of the same charge (positive or negative) must be displaced. The retention factor of the sample ion is kept within the desired range by adjusting the concentration of the competing ion in the eluent.

A specific example may be used to illustrate these principles. A anion exchanger (Anex) is converted to the Cl^- form by passing a solution of sodium chloride through the ion-exchange column. Introduction of a sample containing DNA PO_4^- as an the analyte sets up the exchange equilibrium:

$$\text{xAnex} - \text{Cl}^- + \text{DNA}\,_4^- \rightleftarrows \text{Anex} - \text{DNA PO}_4^- + \text{xCl}^- \tag{3.14}$$

where PO_4^- represents the DNA phosphodiester backbone. Since each DNA molecule has several phosphate groups, an equal number of Cl^- are displaced on the Anex. By adjusting the Cl^- concentration in the eluent to an appropriate value (1 M in some instances) the ratio of Anex–DNA PO_4^- to DNA PO_4^- in solution can be controlled; hence, the retention factor can be kept within desired limits.

The column packing or resin capacity is generally directly proportional to the retention of the sample compounds. Resin capacity is the amount of stationary phase that can actively interact with the sample compound. In the Eqs. described above, the example was based on an insoluble, immobile liquid that was adsorbed on the column packing particles. While early chromatographic work employed particles of this type, modern chromatographic packing materials are usually solid. They consist of a solid substrate on which functional groups are chemically attached. In the case of reverse phase materials, the surface is a neutral organic material usually consisting of an alkyl group 18 carbons in length. The column particle capacity is based on the amount of functional groups that are bonded to the surface of the bead. In some particles the functional groups are attached to substrate in the interior of the bead as well as the outside of the beads. If the sample size is too large to enter the particles, then it is only the stationary phase sites that are accessible to the sample that can be considered to be contributing to the capacity of the packing material.

Ion exchangers behave similarly. Retention increases with ion exchange capacity. Ion exchanger capacity is usually denoted as the number of milliequivalents of ion exchange sites per gram of substrate (mequiv./g).

3.3.5 The Rate Theory of Chromatography

3.3.5.1 General Considerations

Although the treatment given here on the plate theory is simplified, it still shows how a mathematical treatment can be made to describe the chromatographic process. A major limitation of plate theory is that it gives no information on how to adjust conditions to give a low H and thus optimize column efficiency. Plate theory tells the analyst that a longer column will contain a greater number of theoretical plates and give a better separation. Unfortunately, this also increases the time for analysis, causes a greater pressure drop across the column, and can even result in a loss in separation efficiency. Quite obviously, the assumption is false that the eluent flow stops (N times) so that a steady state equilibrium is established with each theoretical plate. Nevertheless, the measured quantities N and H are useful parameters for characterizing chromatography efficiency and are not limited by any of the deficiencies in the plate model. In fact the rate models discussed in this section also enable similar expressions of number of plates and theoretical plate heights [7–9].

The largest shortcoming of the plate model is that it fails to predict the causes of peak broadening. From experience, researchers know that particle size, method of packing, type of stationary phase, solvent viscosity, connections, and mobile phase velocity (flow rate) all contribute to the efficiency of the separation. A famous equation used by chromatographers to measure the effect of these parameters of peak broadening is the van Deemter equation [10].

$$H = A + B/\mu + C\mu \qquad (3.15)$$

where H is the height equivalent of a theoretical plate and μ is the linear flow rate of the liquid through the column with units of mm of column length per second. The linear flow rate is a function of flow rate, column diameter, and the fluid volume of the column. Of course, linear velocity is directly proportional to eluent flow rate and inversely proportional to the square of the column radius. The fluid volume (volume of liquid between the packing material and liquid in the packing pores) of the column depends on how well the particles are packed into the column and the flow through pore volume of the column material. According to the van Deemter equation there are three principle contributions to broadening of a peak signified by the A, B, and C terms.

The A term is called the multipath or eddy diffusion term. In a column the fluid travels many paths at random. Some paths are longer than others and this means that some sample molecules move slower than others while passing through the column. In other words, the flow velocity through a packed column varies widely with flow path through the column resulting in peak broadening. Some sample molecules will travel more rapidly by following open pathways (channeling); others will diffuse into restricted areas and lag behind the zone center (eddy diffusion). In a poorly packed column, flow along the column wall can be quite different than flow through the column bed. These differing flow velocities will cause zone dispersion about the average velocity. Using a homogenous column packing of regular shape and narrow particle size range minimizes multipath diffusion. The column material should be packed very evenly into the column. The A value will normally increase with the average diameter of the packing material. Also, it is important not to leave the column end off and let the column partially dry. This could cause shrinkage of the bed and increase A. The A term is independent of the flow rate, but is instead a function of the column packing.

The molecular longitudinal diffusion B term represents the zone spreading that each sample component exhibits due to diffusion along the column axis (longitudinal diffusion). Diffusion coefficients in aqueous solution are generally low, so the contribution of this term is relatively small unless the retention time is quite long due to a very slow flow rate or a high retention factor. The contribution to band broadening of this term depends on the residence time of the sample compound in the column. Therefore, using a larger eluent flow rate minimizes this term. However, it is only at extremely low flow rates where this term becomes significant – so this term is mostly ignored in liquid chromatography.

The resistance to mass transfer C term relates to the rate at which a sample compound travels to and from the mobile phase and the stationary phase. Mobile phases that have a high viscosity and a low transfer rate would contribute to peak broadening. There is also an additional rate of interaction of the sample material with the stationary phase. For example if the stationary phase is a solid nonpolar surface, the kinetic rate of adsorption and desorption is part of the chromatographic process. If the stationary phase is an ion exchanger, the kinetic rate of exchange process itself is part of the chromatographic process.

Resistance to mass transfer is by far the major contributor to sample zone spreading within the column. This term will be minimized by using a column packing that attains equilibrium as quickly as possible of the sample between the mobile and stationary phases. In other words, the stationary phase sites are completely accessible to the mobile phase. In many cases, a higher eluent temperature will also help.

The contributions of each of the terms in the van Deemter equation are shown in Figure 3.4 where a plot of H as a function of linear flow rate μ is given. The objective is either to minimize h or minimize the time required for separation by using the fastest flow rate that will provide a reasonably low value for H. The A term is usually low for a well-packed column and is not affected by flow rate. As the flow rate increases the B term becomes smaller, but the C term becomes larger. Since B is divided by μ, theoretically this term will go to infinity as μ approaches zero. But the separation time increases with decreasing eluent flow rate and the flow rate at which the B term becomes significant is so low that it would cause impractical high separation times. At some optimum value of μ these terms will balance each other and H and the separation time will be at an optimum. Generally, for most packed columns the volumetric flow rate is about 0.5 to 2.0 mL/min.

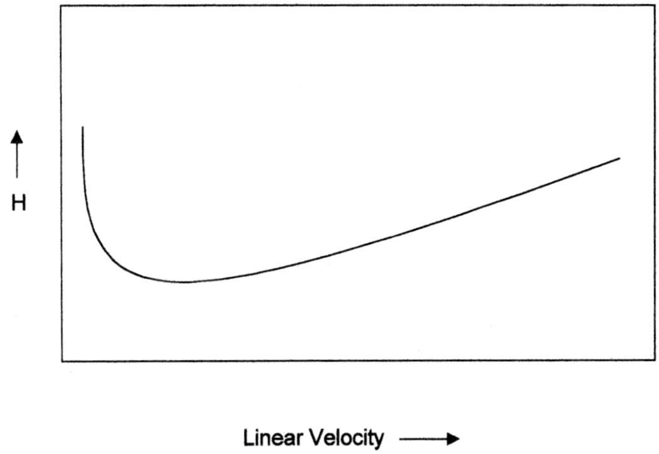

Figure 3.4. The van Deemter plot shows the optimum linear velocity of the eluent to achieve lowest H (height equivalent of a theoretical plate) and the greatest peak resolution. The column length divided by H will give the plate count for a particular column; as H decreases, the total number of theoretical plates increases for the column. At extremely low linear velocities, the column becomes very efficient, but the separation time also increases.

3.3.5.2 Extra Column Effects

We have also noted in the previous chapter that substantial peak broadening may occur *outside* the column. To avoid this the transfer lines connecting the column should be as short as possible. Stagnant areas in the system must be avoided. These can occur, for example, if the connection between two pieces of tubing are not fitted correctly. The detector cell should have a low dead volume. These extra column effects can be measured by injecting a marker (UV absorbing) with no column in the chromatographic system and examining the peak. A system with low extra volume will show a sharp, non tailing peak. A tailing, or broad peak indicates a non optimized system.

3.4 Reverse Phase Column Packing Materials

3.4.1 Types of Materials

There are several classes of materials for general HPLC work. Some of the more commonly used materials are listed in Table 3.3. Of these, reverse phase material is the most common type for HPLC work and is also the most common material for DNA and RNA separations (although most commercial reverse phase column perform poorly for nucleic acid separations). Ion exchange is also used extensively for short, single-stranded DNA separations. This section will discuss reverse phase

Table 3.3. Some common materials used for liquid chromatography

Chromatography	***Material***
Reverse phase	hydrophobic
Normal phase	hydrophilic
Ion exchange	charged ionic sites
Gel permeation	hydrated porous beads
Affinity	selective binding sites

and ion exchange materials synthesis and their use with some specific information directed to materials that are used to separate nucleic acids.

Most liquid chromatography is performed to separate sample molecules that are relatively small i. e. organic acids and alcohols in wine, inorganic ions in drinking water, surfactants in detergents etc. These materials have a molecular weight of only a few hundred. Polymers, proteins and nucleic acid by comparison are extremely large with molecular weights of several thousand units. The requirements to separate large molecules can be much more stringent. Furthermore in the case of nucleic acids, there are at least three additional requirements. One, it is often necessary to separate and analyze molecules that have only one base pair difference. Secondly, the separations are usually size based, not based on chemical differences, and thirdly, because of the high sample load, the separations must be extremely fast and accurate.

These requirements place high standards of performance on the chromatographic material. The material must be highly stable under the conditions used for separation. Temperature stability is needed due to the high temperatures of the mobile phase that are used. The pH used for DNA separation is usually around 7. Silica-based HPLC materials are not as stable at this pH especially with the high aqueous content of the mobile phase and high temperature. Also, high pH (around pH of 10 to 11) column cleaning procedures are often needed to remove sample impurities from the column. Polymer columns stand up to these cleaning procedures better than silica columns. Silica based packing are quite common because historically, they have been quite successful in the separation of small molecules. Almost everyone trained in chromatography prefers silica based HPLC materials and some DNA separations are performed on these materials. However, most separations are performed with polymeric materials.

3.4.2 Polymeric Resins

3.4.2.1 Substrate and Crosslinking

A variety of polymeric substrates can be used in reverse phase synthesis including polymers of esters and amides. Polymer are generally prepared by a two-phase polymerization of vinyl monomers, such as styrene, acrylic esters, vinyl acetate, etc. in a mixture of water, organic crosslinking agent, initiator and a stabilizing agent. Styrene/divinylbenzene copolymers are probably the most common type of polymer reverse phase and ion exchange materials. A schematic of the polymer is shown in Figure 3.5. The resin is made up polystyrene or polyethylstyrene with divinylbenzene added to "crosslink" the various polymer chains in the resin. Crosslinking confers mechanical stability upon the polymer bead by interlocking the polymer chains. Because of the high pressures used by the chromatographic instrument, crosslinking of 55 % to 85 % by weight of crosslinking agent are common for chromatographic polymers. Another reason is that solvents used in chromatography can swell the bead if present or shrink the bead if the organic solvent is taken away (the eluent becomes aqueous). Even with the precaution of using high crosslinking, a polymer always swells at least a little with an organic solvent. Since polymer columns are packed with organic solvents, the polymer can be in a slightly swollen state in the column. Changing the solvent to a completely aqueous mobile phase can remove this solvent and shrink and crack the bed of the column. The best columns can tolerate 100 % aqueous solvents, but it is best to follow manufacturers instructions regarding solvent limits.

3.4.2.2 Porous and Nonporous Resins

The "polymer" resins can be classified either as nonporous or porous. The best DNA separations have been performed on nonporous resins. Nonporous substrates are synthesized by producing a micelle or an emulsion/suspension monomer in suspended in water as organic droplets. The monomers are kept in suspension in the reaction vessel through rapid, uniform stirring. Addition of a free radi-

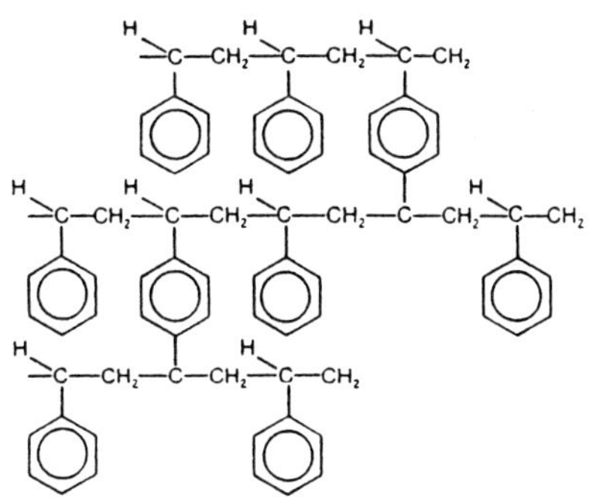

Figure 3.5. Schematic of a styrene–divinylbenzene copolymer. The divinylbenzene "crosslinks" the linear chain of the styrene polymer. A high percentage of divinylbenzene produces a more rigid polymer bead.

cal catalyst such as benzoyl peroxide initiates the polymerization. The mixture is heated with an oil bath or a heating mantle positioned around the reaction vessel. Once the temperature reaches 75°C, benzoylperoxide forms free radicals. This will combine with a monomer to form the start of a polymer with a new free radial on the end. Monomer will add one by one to the chain until either all of the monomers are used or they can no longer reach the end of the polymer chain. The solid beads form rapidly – usually within 30 minutes, but the reaction is continued to be heated for up to 16 hours to ensure that the resin is cured, i. e. make sure that as much as possible of the monomer has been converted into polymer. The resulting beads are uniform and solid and are ready to be cleaned.

The resulting size distribution of the solid beads depends on whether a modified micelle process or an emulsion/suspension process is used. The methods to make polymers usually use a combination of processes to form and maintain the monomer spheres before they are polymerized. Although harder to control and more limiting in the final size of the particle, a micelle process produces the most uniform particles. It is usually unnecessary to perform any additional size classification of the beads when this polymerization process is completed.

Bonn and coworkers [11–13] described the most successful modified micelle process that has been used to prepare packing for DNA separations. The process starts with the formation of a suspension of small polystyrene seeds of uniform size. The seeds are produced by introducing a charge to the surface of the beads during the initial polymerization process. Then the seed is heated or activated to accept transfer of divinylbenzene from an emulsion to the seed particle. The seeds swell to the appropriate size (2 micron) with the fresh monomer and they are polymerized to hard solid beads. After this, the beads are cleaned and the surface is modified to the appropriate polarity by adding an C-18 alkyl group.

The emulsion/suspension polymerization process is much easier to control and can produce a much larger range of particle sized material. Particles as small as 3 micron and as large as 50 micron average particle size can be produced. But the material produced also has a larger particle size distribution and must be sized (usually with air classification) before the resin can be used for chromatographic purposes.

Introducing a detergent into the "monomer mix" and then stirring rapidly in an aqueous buffer produces an emulsion/suspension polymerization. The monomer mix includes the monomer, free radical initiator, and an optional pore modifier (poragen) reagent (described later). When the mixture is heated the polymerization process occurs as described above.

Most materials used for nucleic acid separations are nonporous. The term nonporous can be misleading at times. One reason is because the pore that may be present are too small for the DNA molecules to enter. This is especially true for double- stranded DNA and RNA where the molecules can be very large. As it turns out most polymers have pores as part of the structure. The term nonporous is used for these "slightly porous" polymers because the pores of the particle matrix are not part of the separation process. If only part of the DNA molecule would enter the pore, the kinetic effect would be undesirable because there would be non-uniform or heterogeneous interaction of the DNA with the beads, which in turn would lead to broader peaks. Some column materials are said to have "mega" or very large pores where DNA cannot get trapped. These pores are in effect still part of surface structure and usually do not have a large effect on the selectivity. The material can still be said to be nonporous if the DNA cannot enter the bead matrix and can only interact with the extended surface of the bead. The situation is somewhat different with short (e. g. 20-mer) single-stranded DNA. These molecules are small enough so that they can enter the bead matrix and still maintain sharp chromatographic peaks.

Using various ratios of a diluent or poragen, a solvent which is good for the monomer but poor for the polymer into the monomer mix, can control the pore size of a polymer. The resultant resin bead is spherical and comprised of many hard microspheres interspersed with pores and channels. Again, as with nonporous resins, the polymerization is performed while the monomers are kept in suspension of a polar solvent (usually water). However, the suspended monomer droplets also contain an inert diluent that is a good solvent for the monomers, but not for the material that is already polymerized. The diluents are usually small organic alkane solvents, but could also be small linear polymers of controlled and specific molecular weight. Thus, resin beads are formed that contain pools of diluent distributed throughout the bead matrix. After polymerization is complete, the diluent is washed out of the beads to form the porous structure. The result is rigid, spherical resin beads that have a high surface area – 100 to 400 m^2/g. The exact surface area depends on the type and amount of diluent used. The pore volume of the bead is directly proportional to the amount of diluent used relative to the amount of monomer used. Pore volume can range from 20% to 80% by volume, with 50% pore volume being quite common.

Rather than using an inert solvent to precipitate the copolymer and form the pores, the polymerization may be carried out in the presence of an inert solid agent such as finely divided calcium carbonate to create the voids within the bead. Later, the solid also is extracted (using acid) from the copolymer. Both of these polymerization processes create large (although probably different) inner pores. The average pore diameter can be varied within the range of 20 Å to 500 Å.

The final resin bead structure of a porous resin contains many hard microspheres interspersed with pores and channels. Other terms for porous include macroporous, macroreticular, and megaporous. Because each resin bead is really made up of thousands of smaller beads (something like a popcorn ball), the surface area of macroporous resins is much higher than that of a nonporous resin. A nonporous resin has a (calculated) surface area of less than only a few m^2/g depending on the bead size. However, macroporous resin surface areas range from 25 to as much as 800 m^2/g. Several papers have been published on polymer bead synthesis [14–17].

The chromatography properties of porous polymers can be easily changed by choosing different vinyl monomers or by changing the conditions of the chemical reaction of the polymeric beads including the type and quantity of the poragen and/or the suspension stabilizer. Although styrene based packings remain the most common, more hydrophilic packings based on polyvinyl alcohol, poly (allylmethacrylate), poly hyroxethylmethacrylate), and poly (vinylpyridine) are commercially available [18].

3.4.2.3 Monolith Polymeric Columns

There is another approach to producing polymeric columns that was first introduced by C. Horvath and coworkers for the separation of proteins [19]. The approach starts with a fused silica capillary tubing that has had the walls conditioned to accept attachment of monomers during a polymerization process. Then, a mixture of monomers, poragen and catalyst is filled into the tube and the ends of the capillary are sealed. The mixture is heated starting the polymerization process so that a solid polymer and pore structure is produced inside of the capillary including attachment to the walls. The ends are cut and the capillary is cleaned with solvent making it ready for chromatographic use. This type of column is called a monolith because the polymer structure extends through out the column. No frits are needed to keep the column packing material in place.

A pore structure also extends through out the column allowing solvent to flow through the column and allowing interaction of the sample with the monolith stationary phase. As with conventional materials, the pores for monolith columns are formed from the addition of inert poragens that have been added to the monomer mix. The resulting pore structure of the monolith actually is quite similar to conventional packing materials. The pores that interconnect and extend through the monolith (called the through pores) are analogous to the interstitial spaces between packing particles. The monolith polymer matrix itself also contains a pore structure. These are similar in structure and performance in conventional materials to the pore structure of the packing beads. For DNA chromatography, these monolith matrix pores are quite small and do not allow penetration of the nucleic acids into the resin matrix. Huber and coworkers showed some excellent separations of nucleic acids. These include using the monolith column for mass spectrometry detection [20] and DHPLC [21].

3.4.2.4 Functionalization of the Polymer

The most common method of controlling the surface properties is to cover the surface of the resin bead with chemical functional groups. The most common group for reverse phase groups is to attach C-18 alkyl groups. One reason poly(styrene/divinylbenzene) is so popular is because the aromatic group of the substrate can be reacted in many different ways to attach the alkyl group. An example is shown below.

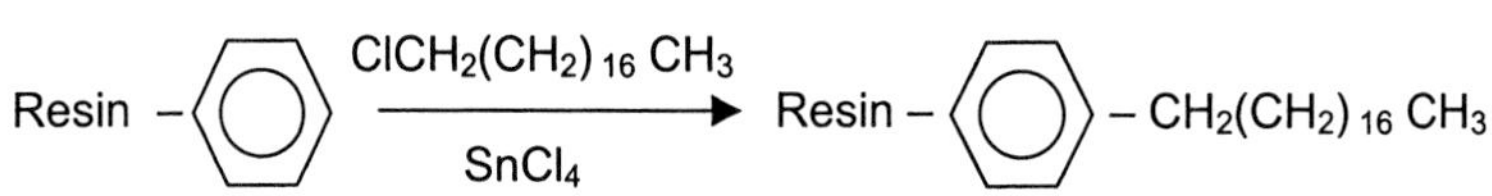

(3.16)

Other functional groups can also be attached including cyano groups, esters, amides and so on. These groups are more polar, but the surface is still hydrophobic and could potentially serve as beads for reverse phase separations. But alkyl groups remain the most popular because resins with this functional group have demonstrated the highest efficiency separations.

A wide variety of resins based on polyacrylate polymers have been produced for use in chromatography. A type known as HEMA, a macroporous copolymer of 2-hydroxyethyl methylmethacrylate and ethylene dimethacrylate, has been used extensively in reverse phase and ion exchange chromatography. It is highly cross-linked to produce a polymeric matrix with high chemical and physical stability. The structure of HEMA is shown in Figure 3.6. Esters are made up of organic acids and alcohols bonded together and they can easily hydrolyze or unbind in aqueous solution (especially at high pH). The highly hindered structure of pivalic acid is one of the most stable and least hydrolyzable esters known, which allows the HEMA stationary phase to be used with a variety of eluents in the pH range 2–12. This is not as good as poly(styrene/divinylbenzene) based substrate, but still considerably better than silica based substrates (described next). The excess hydroxyl groups on the HEMA matrix are quite versatile They can increase the hydrophilicity of this material and improve interaction with polar sample compounds. They can also provide a reaction site upon which more traditional reverse phase groups can be attached.

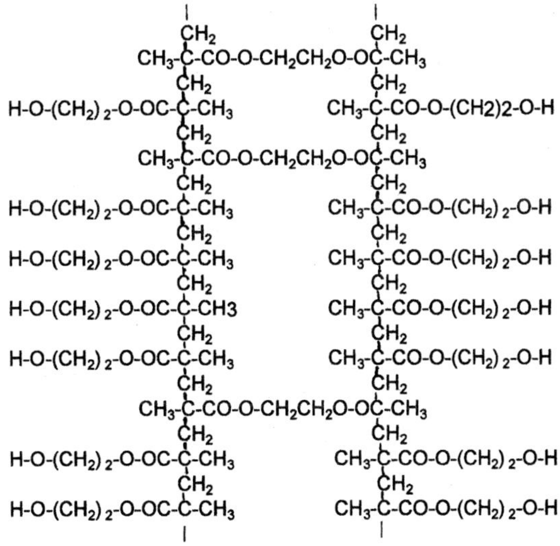

Figure 3.6. The structure of HEMA is shown. Most esters easily hydrolyze or unbind in aqueous solution (especially at high pH). The hindered structure of pivalic acid is one of the most stable and least hydrolyzable esters known, which allows the HEMA stationary phase to be used with a variety of eluents in the pH range 2–12. The hydroxyethyl group allows the chemical attachment of organic functional groups.

3.4.3
Silica-based Materials

3.4.3.1 General Description

As stated before, most recent nucleic acid separations are performed on polymer based substrates because of improved stability and separation power. Still, silica-based substrates are quite common in traditional HPLC. The following is a short description of silica substrates.

The production of silica-based materials for reverse phase chromatography follow a similar path as for polymers. First the substrate is produced and then the materials are reacted to attach a functional group and produce a hydrophobic surface.

The first silica chromatographic materials were prepared using a sol-gel procedure. A hydrosol is initially prepared by the addition of sodium tetraethylsilate to an acidic aqueous solution. Hydrolysis occurs spontaneously followed by condensation to poly silicic acid. An insoluble cake is formed and is eventually dehydrated to form a hard porous material. The caked is milled and air classified to the desired particle size range.

The particles produced by the method described above are irregular and not optimum. Spherical microparticles are more desirable because of their improved performance (as measured by the van Deemter plot). In spite of the fact that spherical silica substrates are more difficult to manufacture most commercial silica based particles are spherical. Spherical silica can be prepared by several methods [22, 23]. One method prepares silica hydrogel beads by emulsification of a silicic acid sol in an immisible organic liquid.

The hydrogel is dispersed into small droplets in the organic liquid and then the temperature, pH, and electrolyte concentration are adjusted to promote solidification. Over time the liquid droplets become more viscous and solidify into bead type particles. The hydrogel beads are then dehydrated to porous spherical silica beads. The particle size and pore diameter are controlled by the size and extent of hydration of the polysilicic acid used to initiate the process. An alternative approach is based on the agglutination of a silica sol by coacervation (to heap up and harvest.)

Urea and formaldehyde are polymerized at low pH in the presence of colloidal poly silicic acid. Coacervated liquid droplets having a spherical shape are formed and settle from the reaction medium. The beads solidify and are harvested and dried. The polymer binder is then burned out to form the pores of the spherical particles [24, 25].

Nonporous silica has also been produced. Early materials, called pellicular, had quite large particle size, and were considered to be less efficient for separations compared to porous silica materials. New material are quite small particle size (as low as 1 micron) and can perform separations quite rapidly. Although very high back pressures can result from these extremely small packings, this can be controlled by using very short columns of perhaps 30 mm [26–28].

3.4.3.2 Functionalization

In order to operate as a reverse phase material, the inorganic silica substrate has a non-polar molecule bonded to its surface The non-polar molecule may be organic polymer or long chain, C-1 to C-24 hydrocarbon groups bound to the inorganic substrate. The silica surface consists of a network of silanol groups and siloxane groups, as shown in Figure 3.7. Bonded phases are prepared by reaction of the surface silanol groups and reactive organosilanes to form siloxane bonds. The organosilane will impart the desired surface effect. A typical reaction is shown below:

$$\text{Substrate} - \underset{|}{\overset{|}{\text{Si}}} - \text{OH} + \text{Cl Si}(\text{CH}_3)_2\text{R} \longrightarrow \text{Substrate} - \underset{|}{\overset{|}{\text{Si}}} - \text{O} - \text{Si}(\text{CH}_3)_2\text{R} \qquad (3.17)$$

where R is the alkyl group of any desired length (usually C-18) or polarity. Initially the reaction may be stoichiometric but as the surface coverage approaches completion, the reaction becomes very slow. Reaction times tend to be long (12–72 hours) with reaction temperatures at about 100 °C.

The reaction is never complete and residual silanol groups always remain. This is the primary reason that silica based materials are more unstable under the conditions used for DNA separations. Since many undesirable chromatographic properties of bonded phases are associated with the presence of these accessible silanol groups, much research continues to reduce their presence or the effect of their presence. One way to accomplish this is by performing a secondary reaction with chlorotrimethylsilane. This reagent is smaller than the original organosilane reagent and can reach the unreacted silanol sites. The replacement of accessible silanol groups in a bonded phase by trimethylsilyl group is generally referred to as endcapping.

Other methods to reduce the effect of silanol groups include polymer encapsulation. This can be prepared by coating the silica substrate with a thin film of prepolymer. This is then crosslinked to form an immobilized skin or the silica surface. The prepolymer may be simply coated on the substrate or it may be bonded through the silanol groups [29–32].

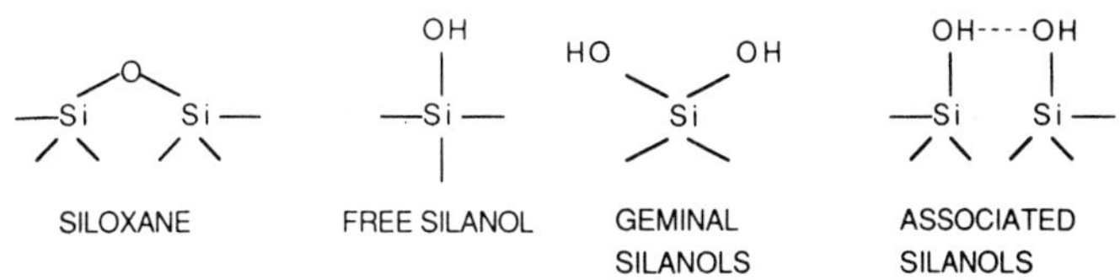

Figure 3.7. Types of silanol groups that may be present on a silica surface. Chemical attachment of functional groups to silica is performed using chloro- or methoxy-silane compounds containing the alkyl or ion exchange group reacted to the silanol sites.

3.5 Reverse Phase Ion Pairing Chromatography

3.5.1 Principles

In order for DNA to be separated by the reverse phase column, it must interact with the stationary phase. The stationary phase is nonionic and hydrophobic. The polymeric DNA molecule is ionic due to the phosphate backbone and, without any treatment, interacts poorly with the stationary phase. The DNA molecules must become non polar in order to develop a chromatographic procedure on the reverse phase column.

Fortunately, there is a reliable and powerful method for achieving this interaction. It is called ion pair chromatography. An additive, an organic cation, as part of the eluent allows the sample nucleic acid to become nonpolar. After injection into the eluent stream, each anionic phosphate group of the nucleic acid is paired with an organic cation. The organic cation is a positively charged ammonium cation that also contains alkyl groups. This results in the formation of an ion pair (N^+A^-) that is more "organic" in nature than the sample ion and could therefore be separated on a reverse phase HPLC column with an organic-aqueous mobile phase such as acetonitrile–water.

In order for the mechanism to operate properly, the eluent must always contain the ion pairing reagent. Therefore if an eluent gradient or acetonitrile is to be used, then the A reservoir (usually low acetonitrile concentration) and the B eluent reservoirs (usually higher acetonitrile concentration) must both the organic ammonium salt.

In the case of DNA chromatography, a fairly small akyl group is used on the organic cation. The most common one is triethylammonium cation. To prepare this, acetic acid is added to triethylamine to make a final solution pH of 7. At this pH, the triethylamine is protonated to the ammonium cation:

$$N(CH_2CH_3)_3 + HAc \rightarrow H\,N^+(CH_2CH_3)_3 + Ac^- \qquad (3.18)$$

where Ac^- denotes acetate. The separation process is shown in Figure 3.8. The mobile phase is pumped through a reversed phase HPLC column until equilibrium is attained. Then a sample is injected and the anions from the DNA combine with the opposite charged ammonium cations. The alkyl groups cause the whole molecule of DNA to become hydrophobic and the molecule adsorbs to the stationary phase. A gradient of acetonitrile is started. First, the smaller DNA molecules are desorbed and travel down the column and to the detector. Then, as the organic solvent content of the eluent is increased, larger and larger DNA molecules are desorbed and also travel down the column. Finally after all the fragments of interest are eluted, the column is cleaned of residual material with a final gradient concentration of high acetonitrile. After this, the eluent is returned to original eluent concentrations and the column is conditioned for the next run.

The injection solvent (the solvent containing the sample) should not be stronger than the eluent otherwise it will induce a premature start to the chromatographic

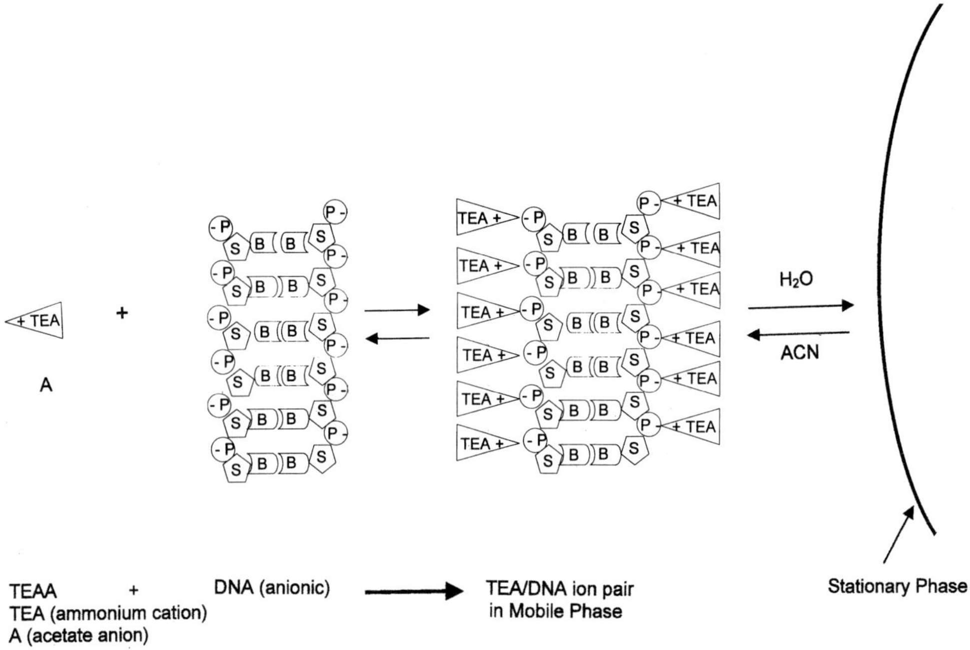

Figure 3.8. The ion pairing reverse phase process. TEA denotes triethylammonium, A acetate, P phosphate, S sugar, and B base. The ion pairing reagent is prepared by mixing triethylamine with acetic acid to form triethyammonium acetate (TEAA). The negatively charged nucleic acid is combined with the positively charged triethylammonium ion to form the neutral ion pair. The ion pair can interact with the surface of the column packing (the stationary phase). As the water content of the mobile phase increases, the equilibrium or adsorption to the neutral non porous surface increases. As the acetonitrile concentration of the mobile phase increases, the equilibrium or adsorption to the surface decreases.

process. For example, if the start of a separation is performed at 10 % acetonitrile (40 % of B eluent that is 25 % acetonitrile), but the sample acetonitrile solvent is 15 %, then some of the sample will start traveling down the column before the gradient is initiated. This will lead to broad non reproducible peaks or may even wash the sample through the column and completely prevent separation. Some instrumentation includes an injection needle wash. Likewise, if the needle or loop wash solvent contains organic solvent then it may also wash the sample down the column prematurely.

Double-stranded DNA is separated on the basis of size. The sequence of the fragment does not contribute to the separation. This is very unusual in liquid chromatography where separations are usually based on the chemical nature of the sample molecules. Single-stranded DNA and RNA also separate on the basis of size, but the separations are not completely sequence independent. Since these molecules contain only half as many phosphodiester groups as double-stranded DNA, they also contain only half as many nonpolar ion pairing molecules. Therefore, the polarity effects due to the single-stranded nucleic acid backbone have a larger effect in retention. Retention times of single-stranded nucleic acids are lower and less acetonitrile is needed to elute them from the column. Fragments containing more C and G nucleotides will elute earlier in the chromatogram than fragments that contain more A and T nucleotides. The effect is even greater if the C and G nucleotides are on the end of the fragment. This sequence dependent effect can be lessened for single-stranded DNA if a more nonpolar (hydrophobic) ion pairing reagent such as tetrabutylammonium bromide is used instead the usual triethylammonium acetate [33].

The following experimental parameters can be adjusted to obtain satisfactory conditions for a separation:

1. Type of stationary phase. More contact surface will increase retention. A more hydrophobic surface will increase retention.
2. Type of pairing reagent. A more nonpolar reagent shifts the equilibrium towards the surface association.
3. Concentration of pairing reagent. A higher concentration shifts the equilibrium towards the surface association.
4. Type and concentration of organic solvent (organic modifier). Increasing concentrations shift the equilibrium away from the surface.

The mechanism of what we call "ion-pair chromatography" has been the subject of a considerable amount of investigation. Horvath *et al.* demonstrated the practicality of this approach [34]. They proposed an ion-pair mechanism and gave a number of ion-pair formation constants. Kraak, Jonker and Huber used anionic surfactants in conjunction with a bonded-phase silica column and an organic-aqueous mobile phase for the separation of amino acids [35]. A comprehensive study was made of the parameters, including the generation of gradients. Kissinger argued that an ion-pair mechanism is incorrect [36]. The pairing reagent partitions strongly onto the stationary phase, modifying its surface charge. This implies an ion-exchange mechanism. This interpretation would appear to be valid when the pairing reagent is very strongly adsorbed on the stationary phase surface. Bidlingmeyer and coworker also argued that the mechanism was a combination of ion exchange and adsorption [37].

In DNA chromatography, it appears that the separation is based on a desorption of the ion-pair from the column as the acetonitrile concentration reaches a certain concentration. The reasons for stating this are based on certain observations including:

1. When the mobile phase composition is varied for isocratic elution (no gradient used) it is found that the sample never migrates normally, but is either completely retained or completely eluted with the void volume. In other words, depending on the amount of acetonitrile in the eluent, either the DNA completely sticks to the column or it does not interact at all and travels with the injection solvent plug (into the mobile phase) and through the column.
2. When the column length is varied in gradient elution, the resulting chromatogram changes very little. A 3 cm column will give virtually the same separation

as a 10 cm column. There is little difference in the retention times and total separation time when sample are run under identical conditions with columns of two different lengths.

3. A particular size fragment of DNA will elute at a particular concentration of acetonitrile regardless of slope of the gradient (how quickly the gradient is generated).

Figure 3.9 illustrates that both desorption and partitioning can occur under DNA Chromatography conditions. For the purpose of this illustration, it is assumed that there is only one fragment of 200 bp in the sample. Figure 3.9a shows that under steep gradient conditions, the fragment is first adsorbed on the column, but is stripped and quickly begins to travel at the same velocity as the eluent through the column. This is known because the conditions between complete adsorption from a single bulk equilibration and complete desorption have been found to be only a couple of percent of acetonitrile. A steep gradient will go through this change rapidly and cause the fragment to completely dissolve in the mobile phase. At that point, the fragment is moving at the same velocity as the mobile phase. The Figure shows that for steep gradients, this happens quite quickly in the chromatogram and the fragment can interact only with the first narrow part of the column. This interaction is called partitioning. The sample fragment is moving back and forth from the stationary phase and mobile phase forming the basis for the separation. It is likely that even with a very sharp gradient, some small amount of partitioning of the DNA is occurring at the first part of the column. Therefore, it is also likely that the basis for separation of the fragments is simply the difference in acetonitrile concentration needed to desorb a particular size fragment. Under these conditions a column that is half the size or a quarter of the size will give almost an identical separation to the full size column.

Figure 3.9b shows that as the steepness of the gradient is lowered, it takes longer for the fragment to completely desorb into the mobile phase. Therefore, the transition for being completely adsorbed to completely desorbed covers a longer portion of the column. As stated above, partitioning occurs over this part of the column. The partitioning mechanism of the separation is being “turned on” as the steepness in the gradient is being lowered. More and more of the column is being used. However, a less steep gradient covers a smaller fragment size range or has a longer separation time.

Finally, Figure 3.9c shows that when a column is run under very shallow gradient conditions or uniform eluent conditions (isocratic elution) the full length of the column is being used for the separation. Under these conditions, the highest resolution of DNA fragments can occur. But there is a penalty. The conditions for separation must be chosen exactly. Unless the right acetonitrile concentration is chosen, the fragment will either stick to the top of the column and not move or it will travel through the column quickly with the void volume. It is only if the conditions are exactly right will the fragments of interest be separated. Therefore, under most situations, a gradient is used. Fortunately, the penalty of lower resolution is not great when using the high performing columns that are available today.

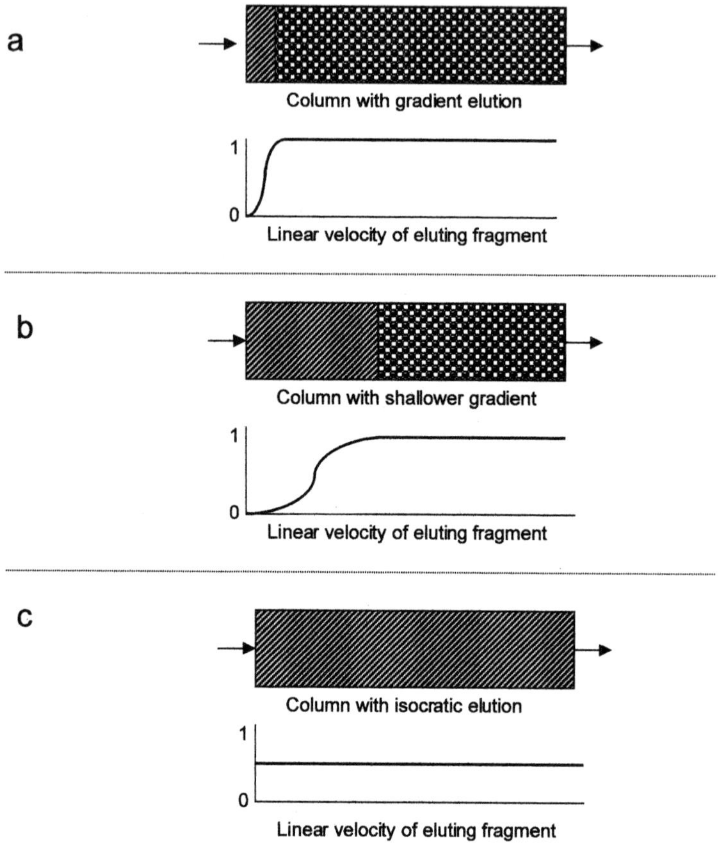

Figure 3.9. Both desorption and partitioning can occur under DNA Chromatography conditions. a) When an eluent gradient is used, only the top portion of the column is used (denoted by slanted lines) and each fragment is effectively desorbed from the column when the desorption acetonitrile concentration is reached for each fragment. At the point that the fragment is no longer is interacting with the stationary phase (denoted by circles) and it is traveling through the column at the same linear velocity as the eluent. The plot below the column shows the linear velocity of the fragment as it travels through the column. b) Lowering the slope of the gradient will allow more of the column to be used in the separation process because the conditions for complete desorption for a particular fragment are reached later. c) In isocratic elution, the fragment is interacting with the column and eluent, and the fragment is traveling through the column at a constant velocity but is also lower than the linear velocity of the eluent. The velocity of the fragment will approach the velocity of the eluent remaining constant as the isocratic amount of the acetonitrile in the eluent is increased.

3.5.2
Temperature Modes of DNA Chromatography

In the previous chapter we discussed the importance of the oven in the DNA chromatography apparatus. The oven controls the temperature of the fluid entering the separation column. This includes not only the eluent but the sample that is injected into the system. Temperature of the fluid can be thought of as an additional reagent used in the separation and analysis of nucleic acids. This is best described in the three modes of operation of analysis: Non Denaturing Mode, Partially Denaturing Mode, and Fully Denaturing Mode.

1. Non Denaturing Mode. The breaking of hydrogen bonds of double-stranded DNA through increasing the temperature is called melting. The temperature at which this occurs depends on the strength of the hydrogen bonding and the environment around the DNA. Higher salt and buffer content will raise the melting temperature. Also, DNA adsorbed onto a solid surface, such as column packing material, will require higher temperature to melt the DNA. Conversely, the presence of an organic solvent such as acetonitrile will lower the melting temperature. The effect is increased with increasing concentrations of the solvent. In the non denaturing mode, high enough temperature is chosen to lower eluent viscosity (and therefore column back pressure), but not so high that denaturing occurs. The normal temperature of operation is 50 °C.
2. Partially Denaturing Mode. The use of partially denaturing mode is discussed in Chapter 4 on DHPLC. Mutations are detected through a heteroduplex detection process. Under this mode, differences in melting of double-stranded DNA are detected because increased melting decreases the retention time of the fragment. The temperature is chosen depending on the sequence and the melting of the fragment being analyzed. Typically, oven temperatures between 54 and 72 °C are used. The method has been called a difference detecting engine because it detects the presence of a heteroduplex regardless of the sequence being studied.
3. Fully Denaturing Mode. Single-stranded DNA and RNA may contain secondary structure due to complementary sequences within its own fragment. If portions of sequences are complementary the fragments may fold back on themselves through the formation of intra-molecular hydrogen bonds giving several different possible structures for the same fragment. The presence of these secondary structures is uncertain and non reproducible and will lead to peak broadening if they occur. Increasing the temperature can break up the hydrogen bonding and therefore reduce some or perhaps most of the secondary structure [38]. Figures 3.10a and 3.10b show the separation of an RNA ladder at 40 and 75 °C. The peaks of the RNA become sharper and more uniform with increasing oven temperature. It is not certain that temperature will always reduce secondary structure. But in many cases it will.

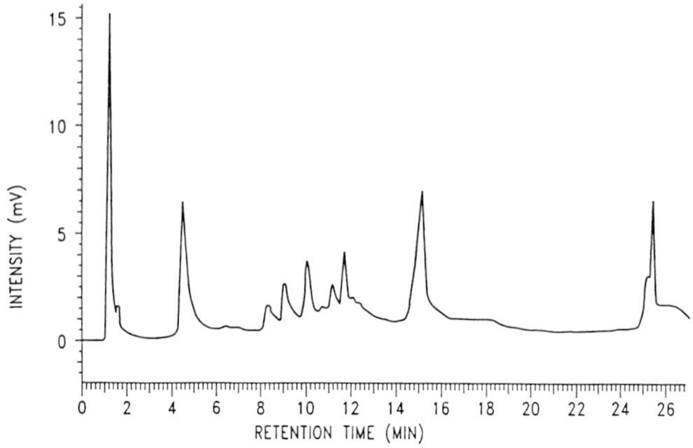

Figure 3.10. (a) Chromatogram of RNA ladder at 40 °C. RNA ladder (Cat. No. 15623010, Life Technologies) has nucleotide lengths of 155, 280, 400, 530, 780, 1280, 1520, and 1770 bases. Buffer A: 1 M TEAA, pH 7.0, buffer B: 1 M TEAA, pH 7.0 with 25 % v/v acetonitrile, gradient 0.0 min 38 % B, 1.0 min 40 %, 16 min 60 %, 22 min 66 %, 22.5 min 70 %, and 23 min 100 %. (from Ref. [38] with permission).

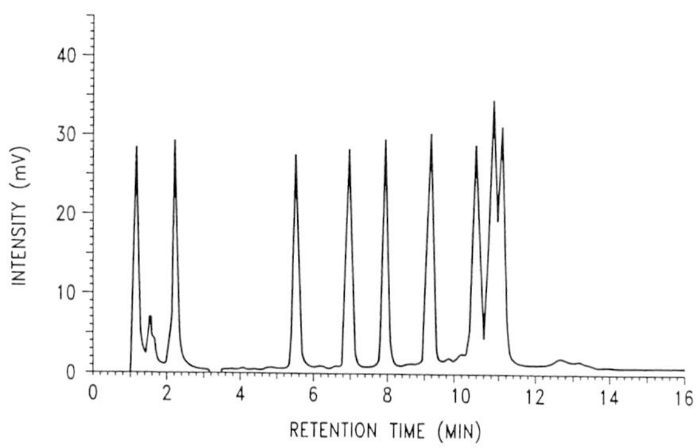

Figure 3.10. (b). Same conditions as Figure 3.10a except chromatogram is at 75 °C. It is not certain that temperature will always reduce secondary structure but in many case it will (see Chapter 7).

3.5.3 Effect of Metal Contamination

Günther Bonn and his colleagues, Christian Huber and Peter Oefner had developed the world's leading chromatographic method for separating double-stranded and single-stranded polynucleotides [11–13]. In the course to commercialize their technology, a degradation effect was discovered after the column had experienced medium to longer term use [39–42]. The DNASep® column would work for a while and then would start giving either split peaks for each fragment eluting from the column or stop giving peaks at all. At first, the packing procedure was extensively studied because split peaks means that the chromatographic bed has become cracked or disrupted. This was not unexpected since polymer beds are prone to this kind of behavior if they are not packed properly. The packing procedure was studied to rule out this possibility of unstable packed column beds.

Then, the column was put into a standard HPLC system that destroyed the separation slowly and somewhat controllably. In one set of experiments, a new column showed excellent separation of a pUC 18 *Hae* III digest (Figure 3.11a). As the column was used, a degradation effect was observed as a loss of resolution for base pairs greater than 200 (Figure 3.11b). As the degradation continued, increasingly shorter fragments of DNA were affected. Many of the peaks were split or doublets (Figure 3.11c). Eventually, the DNA did not elute at all from the system. As such, the degradation or decreasing resolution appeared to be a function of the length of the polynucleotide fragment being separated. Although several causes for the behavior were discussed including column contamination, the researchers first examined different procedures for packing the column or for performing the separation. They realized that the DNA molecules were several magnitudes larger in size than those conventionally separated by reverse phase ion pair liquid chromatography. They suspected that the hydrodynamic flow through the column was adequate for short DNA fragments, but was being disrupted for larger fragments. In other words, perhaps the longest fragments were being partially sheared. However, they were unable to identify packing materials or separation conditions that caused the same degradation pattern. Furthermore, if shearing occurred, the process would be random and the chromatogram pattern would not be the same.

Although the researchers could not conceive a mechanism by which chemical contamination could produce these unusual results, they nevertheless examined contamination of one or more of the various "pure" reagents employed in liquid chromatography. After testing each of the reagents for contamination, they determined that this was not the source of the problem. This is not surprising, since the mobile phases used are not corrosive.

The cause of the split peaks for the larger DNA fragments was discovered when the attempts were made to clean the column. Clean up with organic solvents did not improve performance. However, subsequent clean-up of the column with injections of tetrasodium ethylenediaminetetraacetic acid (EDTA), a metal-chelating agent, largely restored chromatographic resolution. It appeared that the chelating reagent either passivated metals in the system or removed metals form the system so that they could not bind with the DNA. It is still not certain why, under certain circumstance, the contamination affects only large fragments at first. However, experiments to remove corroded HPLC parts and prevention of corrosion all helped improve performance.

Adding small amounts (i. e., 0.1 mM) of tetrasodium EDTA to the mobile phase can be done without significant changes to the chromatography. The most significant sources of metal ions are HPLC components containing fritted filters made of stainless steel. Fritted filter components are used in mobile phase filters, check valve filters, helium spargers, mobile phase mixers, inline filters, column frits, and other parts of the HPLC. Frits are commonly located at each end of a separation column in order to contain the particulate packing material within the column. The frit at the head of a column also serves to trap particulate material.

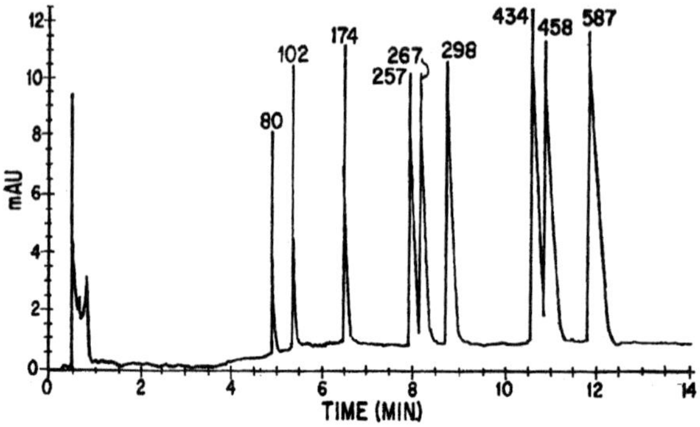

Figure 3.11a. A series of three chromatograms are shown separation is shown illustrating the degradation effect on peak separation of double-stranded DNA, pUC18 *Hae* III. The degradation occurs first on larger fragments. Smaller fragments are eventually degraded as well (from Ref. [40] with permission). (a) This first chromatogram uses a new DNASep® column, Buffer A: 1 M TEAA, pH 7.0, buffer B: 1 M TEAA, pH 7.0 with 25% v/v acetonitrile, gradient, 0 min 35% B, 3 min 55% B, 10 min 65% B, and 12 min 100% B.

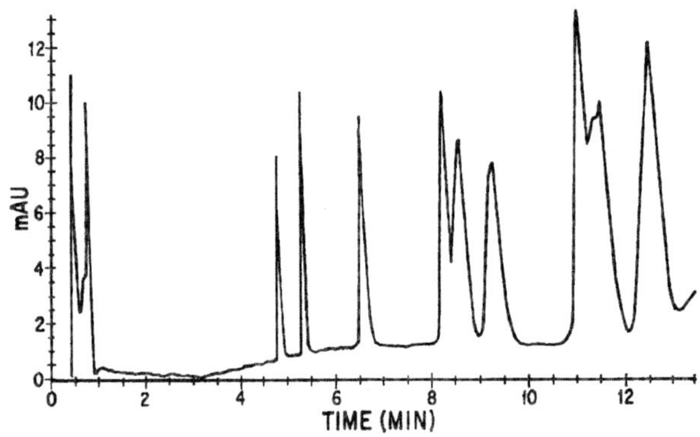

Figure 3.11b. After several minutes of use, the column shows the initial degradation of the HPLC separation with fragments greater than about 200 bp affected

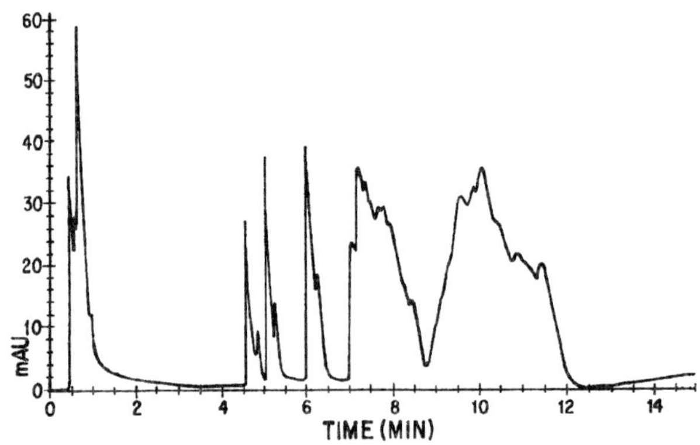

Figure 3.11c. After several more minutes of use, the column shows the greater degradation of the HPLC separation. Now all fragments are affected.

Trapped particulate materials can be metal ions released from another part of the liquid chromatography system. The large surface area associated with any particular fritted component can contribute to faster solubilization of metals and release of ions. Thus, the contamination from a fritted component can arise in at least two ways. First, the component can be a source of ionic material. Second, it can be a means for trapping ionic material.

Metal ion contamination such as colloidal iron can be released from frits, travel to other parts of the HPLC and then be trapped. These types of contaminants will interfere with DNA in solution or after having been released and trapped on a critical component of the HPLC such as the column, an inline filter in front of the detector, or at a back pressure device located after the detector.

Subsequent experiments showed that even if titanium or PEEK fluid path components are used, then some treatment was necessary before the components could be used. Although an improvement, initial use of titanium frits did not give consistent results. Treatment of the frits with dilute nitric acid and then with a chelating agent did improve the performance of the instrument. Similarly, as shown in the examples, PEEK frits were not consistently suitable for DNA Chromatography, but acid treatment did improve their performance. Finally, degassing the fluid before it enters the liquid chromatography system removes the oxygen. This process will inhibit the oxidation and production of metal ions in stainless steel or titanium or other tubing containing iron.

All of these improvements are incorporated into commercial instruments designed for DNA Chromatography. It is important to follow the manufacturers procedures in keeping the eluents degassed and clean the column regularly. It is also important to follow preventive maintenance procedures and check filters for signs of colored deposits. Appendix 2 shows a procedure used to remove corrosion from a system. This procedure should be used only when the column is removed. Since double-stranded DNA is more susceptible to contamination use of the precautions of the method and system are much more critical than when the system is used to separate single-stranded DNA.

3.6 Ion Exchange Materials and Separation Mechanism

3.6.1 Polymer-based Anion Exchangers (Anex)

Ion exchange chromatography has had a long history. Glass, clays, minerals, and natural occurring organic substances have ion exchange properties and have been exploited for commercial purposes. But heavy use of ion change did not really start until the 1940s with the development of synthetic polymeric ion exchangers. Ion exchange chromatography is used extensively to purify useable quantities of materials, especially in the production of rare earth metals. These metals are similar chemically and are very difficult to purify by any other method. Ion exchange is also popular to isolate preparative amounts of proteins because the activity of the protein is retained with this procedure.

Analytical uses of ion exchange include the analysis of small ions such as chloride, sulfate and others in drinking water, power plant water, plating baths, food and beverage products. Anion exchangers with diethylaminoethyl groups were used by Y. Kato et al. to separate polynucleotide fragments [43]. The most important application of anion exchange for DNA is the analysis and purification of short single-stranded DNA.

An important disadvantage of anion exchange separations of double-stranded and single-stranded nucleic acids is the differing retention behavior of GC- and AT-base pairs. This effect makes separation according to molecular size impossible. W. Bloch demonstrated that, to a certain extent, length relevant separation of double-stranded DNA fragments was possible on nonporous anion exchangers with tetramethylammonium chloride (TMAC) containing mobile phases [44]. Another important drawback of the anion exchange methodology is the necessity to use salts and buffers for elution, thus making subsequent recovery of the DNA molecule fractions more difficult.

Ion exchangers can be either anion exchange or cation exchange. There is one phosphate group on each nucleotide each contributing one negative charge to the total molecular charge. Since nucleic acids are negatively charged, they are separated with an anion exchanger. Discussions in this book on ion exchangers are limited to anion exchangers.

A anion exchanger is a solid particulate material with positively charged functional groups arranged in a manner which interact with ions in the surrounding liquid phase. Anex materials are prepared in a similar way as the reverse phase materials described earlier in this chapter. First a substrate or bead is prepared and then the surface is reacted to attached the ion exchange site. The substrate materials are prepared the same way and then the surfaces are reacted to form the ion exchange sites. As before, the substrates may be polymer or silica-based. A schematic of an anex material is shown below:

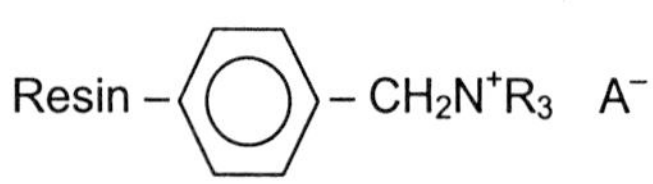

(3.19)

Resin denotes polymer, R denotes an alkyl group and A denotes an exchangeable anion. The nitrogen has four bonded alkyl groups giving it a permanent positive charge regardless of the eluent pH. This is sometime called a strong base anion exchanger. Other anion exchangers have been used where there are only three alkyl groups bonded to the nitrogen atom:

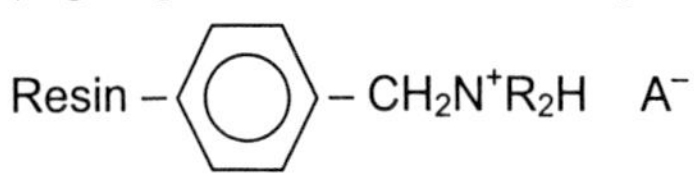

(3.20)

where H is a hydronium cation. In many cases, the R group is an ethyl group and the resin is called a DEAE (diethyl anion exchanger). In this case, the nitrogen must be protonated before it can operate as an anion exchanger and these resins are used in low pH eluents.

A wide variety of resins based on polyacrylate polymers has been produced for use in chromatography. A type described earlier is HEMA, a macroporous copolymer of 2-hydroxyethyl methylmethacrylate and ethylene dimethacrylate, has been used extensively in ion exchange chromatography. It is extensively crosslinked to produce a polymeric matrix with high chemical and physical stability. The anion exchange site is attached through the hydroxyethyl group.

Porous and nonporous resins are used in anion exchange chromatography. The Dionex Co., Sunnyvale, CA, has developed a number of latex-coated nonporous materials. Several of their anion exchange resins have a surface layer of quaternary ammonium latex on a surface-sulfonated substrate. The schematic for these anion exchangers is:

$$\text{Resin} - SO_3^- \; N^+ R_3 - \text{latex} - N^+ \;{}_3 \qquad (3.21)$$

Some of the anion exchange sites in the latex are used to electostatically attach the latex to the resin substrate. This leaves the outside anion exchange sites of the latex free to be used in anion exchange chromatography.

Whether a particular material can be used for the separation of nucleic acids depends on a number of factors including pore size, ion exchange capacity, stability, etc. It is important to follow the manufacturers recommendations on what materials are suitable.

3.6.2
Silica-based Anion Exchangers

Silica-based cation exchangers are generally prepared by reacting silica particles with an appropriate chlorosilane or methoxysilane. A common type of silica anex has the structure:

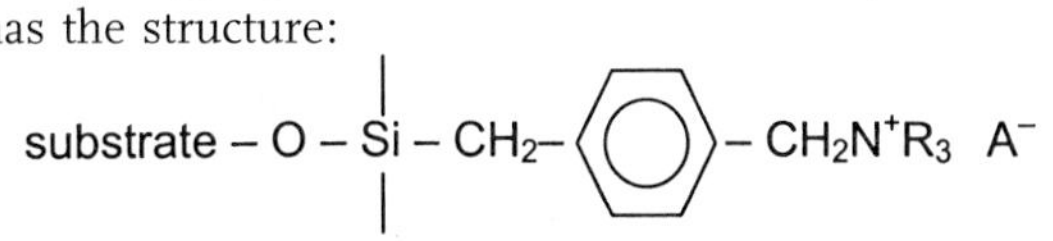

(3.22)

Compared to organic polymers, silica-based ion exchangers have the advantages of higher chromatographic efficiency and greater mechanical stability. In general, no problems due to swelling or shrinking are encountered, even if an organic solvent is added to the eluent. A disadvantage of silica materials is their limited stability at lower pH values and especially in alkaline solutions. A fairly narrow pH range of 2 to 8 is recommended.

Silica-based anion exchangers are available from a number of manufacturers. Typical trade names include Vydac (Separations Group, Hesperia, CA, U. S. A.), TSK Gel (Toyo Soda, Tokyo, Japan) and Nucleosil (Machery & Nagel, Düren, Germany). As described before there are two distinct types of silica resins. Totally porous resins have a quaternary ammonium functional group chemically attached. Their particle size is in the range of 3 μm to 10 μm with typical exchange capacities of 0.1 to 0.3 meq/g. Nonporous, sometimes called pellicular, materials have a larger particle size and are covered with a thin layer of a polymer with quaternary ammonium groups. For example, Zipax SAX is covered with a layer of lauryl methacrylate 1 μm to 3 μm thick.

3.6.3 Basis for Separation

The basis for separation in ion exchange chromatography lies in differences in the exchange equilibrium between the various sample anions and the eluent ion. To perform a separation, eluent is first pumped through the system until equilibrium is obtained, as evidenced by a stable baseline. At this point all of the ion exchange sites are in the eluent form. For example, if sodium chloride eluent is equilibrated with the column, then all of the anion exchange sites will have eluent chloride anion E^- associated with them. Another way of saying this is the anex is in the E^- form or Cl^- form. The time needed for to achieve may vary from a couple of minutes to an hour or longer, depending on the type of resin and eluent that is used.

An analytical sample can be injected into the system as soon as a steady baseline has been obtained. A sample containing fragments F_1, F_2, F_3,. . . F_i each undergo ion-exchange with the exchange sites near the top of the chromatographic column.

$$F_1 + x\text{Resin} - N^+R_3\,E^- \rightleftarrows x\ \text{Resin} - N^+R_3\,F_1 + xE^- \tag{3.23a}$$

$$F_2 + y\text{Resin} - N^+R_3\,E^- \rightleftarrows y\ \text{Resin} - N^+R_3\,F_2 + yE^- \tag{3.23b}$$

$$F_3 + z\text{Resin} - N^+R_3\,E^- \rightleftarrows z\ \text{Resin} - N^+R_3\,F_3 + zE^- \tag{3.23c}$$

etc.

Each fragment contains one phosphate group or one negative charge for every nucleotide of the fragment. Every time a fragment exchanges with the anex, x number of eluent ions are displaced.

The general principles for separation are perhaps best illustrated by a specific example. Suppose that a 20-mer oligonucleotide and 19-mer failure sequence are to be separated on an anion-exchange column with a 1.0 M NaCl eluent mobile phase.

In the column equilibration step, the column packed with solid anion-exchange particles (designated as N^+Cl^-) is washed continuously with the NaCl eluent to convert the ion exchanger completely to the -Cl^- form. In the sample injection step a small volume of sample is injected into the ion-exchange column. An ion exchange equilibrium occurs in a fairly narrow zone near the top of the column.

$$19\ N^+Cl^- + 19F^- \rightleftarrows 19\ N^+F^- + 19\ Cl^- \tag{3.24a}$$

$$20\ N^+Cl^- + 20F^- \rightleftarrows 20\ N^+F^- + 20\ Cl^- \tag{3.24b}$$

where the 19-mer failure sequence is denoted as $19F^-$ and the 20-mer fragment is denoted as $20F^-$. Each fragment displaces 19 and 20 Cl^- anions, respectively, when the fragment exchanges with the anion exchanger.

In isocratic elution a constant eluent concentration is pumped through the column. Using the example of a 1 M NaCl eluent through the column results in multiple ion-exchange equilibria along the column in which the sample fragments ($19F^-$ and $20F^-$) and eluent ions (Cl^-) compete for ion-exchange sites next to the N^+ groups. The net result is that both $19F^-$ and $20F^-$ move down the column.

Because $20F^-$ has a greater affinity for the N^+ sites than $19F^-$, the $20F^-$ moves at a slower rate. Due to their differences in rate of movement, $20F^-$ and $19F^-$ are gradually resolved into separate zones or bands.

However, isocratic elution does not work well for the ion exchange of nucleic acids. The ion exchange binding increases rapidly with increasing length of the nucleic acid. A gradient must usually be used to elute nucleic acids of different lengths in the same chromatogram. The ion exchange mechanism is similar to that described in Figure 3.9 with fragment interaction with the column becoming less as the steepness of the gradient is increased.

References

1. M. Tswett, Adsorption Analysis and Chromatographic Methods. Application to the Chemistry of Chlorophyll, Berichte der Deutschen botanischen Gesellschft 24, 385 Ber. Deut. Botan. Ges., 24, 384, 1906.
2. C. F. Poole and S. K. Poole, Chromatography Today, Elsevier: New York, 1991.
3. L. R. Snyder and J. J. Kirkland, Introduction to Modern Liquid Chromatography, 2nd ed, Wiley–Interscience: New York, 1979.
4. Chromatography, 5th ed. Part A: Fundamentals and Techniques, edited by H. Heftmann, Elsevier: Amsterdam, 1992.
5. L. S. Ettre and V. R. Meyer, Two symposia, when HPLC was young, LC/GC Magazine, July 2000.
6. C. Horvath in 75 Years of Chromatography – A Historical Dialogue, edited by L. S. Ettre and A. Zlatkis, eds., Elsevier: Amsterdam, 1974, pp. 704–714.
7. J. C. Giddings "Dynamics of Chromatography", Marcel Dekker: New York 1965.
8. E. Katz, K. L. Ogan, and R. P. W. Scott, Peak dispersion and mobile phase velocity in liquid chromatography: the pertinent relationship for porous silica, J. Chromatogr. 270, 51, 1983.
9. E. Grushka and E. Kikta, Chromatographic broadening technique of liquid diffusivity measurements, J. Phys. Chem., 79, 2199, 1975.
10. J. J. van Deemter, E. J. Zuiderweg, and A. Klinkenberg, Longitutional diffusion and resistence to mass transfer as causes of nonideality in chromatography, Chem. Eng. Sci., 5, 271, 1956.
11. C. G. Huber, P. J. Oefner, E. Preuss, and G. K. Bonn, High-resolution liquid chromatography of DNA fragments on highly cross-linked poly(styrene-divinylbenzene) particles, Nuceic Acids Res. 21, 1061, 1993.
12. C. G. Huber, P. J. Oefner, and G. K. Bonn, Rapid analysis of biopolymers on modified non-porous polystryrene – divinylbenzene particles. Chromatographia, 37, 653, 1993.
13. G. Bonn, C. Huber, and P. Oefner, Nucleic acid separation on alkylated nonporous polymer beads, U. S. Pat. 5,585,236, 1996.
14. F. Nevejans and M. Verzele, Swelling propensity (SP factor) of semi-rigid chromatographic packing materials, J. Chromatogr. 350, 145, 1985.
15. F. Nevejans and M. Verzele, On the structure and chromatographic behavior of polystyrene phases, J. Chromatogr. 406, 325, 1987.
16. F. Nevejans and M. Verzele, Porous polystyrene packings: characterization through pore structure determination, Chromatographia 20, 173, 1985.
17. H. W. Stuurman, J. Kohler, S. O. Jansson, and A. Litzen, Characterization of some commercial poly9styrene-divinylbenzene) copolymers for reversed-phase HPLC, Chromatographia, 23 (1987) 341
18. F. M. Rabel, Use and maintenance of microparticle high performance liquid chromatography columns, J. Chromatogr. Sci., 18 (1980) 394.
19. I. Gurev, X Huang, and C. Horvath, Capillary columns with in situ formed monolithic packing for micro high-performance liquid chromatography and capillary electrochromatography, J. Chromatogr., 855, 272, 1999.
20. A. Premstaller, H. Oberacher, and C. G. Huber, High performance liquid chroma-

tography-electrospray ionization mass spectrometry of single- and double- stranded nucleic acids using monolithic capillary columns, Anal. Chem. 72, 4386, 2000.

21. C. G. Huber, A. Premstaller, W. Xiao, H. Oberacher, G. K. Bonn, and P. J. Oefner, Mutation detection by capillary denaturing high-performance liquid chromatography using monolithic columns. J. Biochem. Biophys. Methods 47, 5, 2001.
22. K. K. Unger "Porous Silica" Elsevier, Amsterdam, 1979.
23. K. K. Unger, J. N. Kinkel, B. Anspach, and H. Giesche, Evaluation of advanced silica packings for the separation of biopolymers by high-performance liquid chromatography, I. Design and properties of parent silicas, J. Chromatogr., 296, 3, 1984.
24. N. D. Danielson, and J. J. Kirkland, Synthesis and characterization of 2 μm wide-pore silica microspheres as column packing for the reversed-phase liquid chromatographic separation of peptides and proteins, Anal. Chem. 59, 2501, 1987.
25. R. W. Stout, G B. Cox, and T. J. Odiorne, Surface treatment and porisity control of porous silica microspheres, Chromatographia, 24, 602, 1987.
26. T. J. Barder, P. J. Wahlman, C. Thrall, and P. D. DuBios, Fast Chromatography and Nonporous Silica, LC–GC, 15, 918-926, 1997.
27. H. Chen and C. Horvath, High-Speed High-Performance Liquid Chromatography of Peptides and Proteins, J. Chromatogr. 705, 3, 1995.
28. W. Stober, A. Fink, and E. Bohn, Controlled growth of monodisperse silica spheres in the micron size range, J. of Coll. and Interface Science, 26, 62, 1968.
29. H. Engelhardt, H. Loew, W. Eberhardt, and M Mauss, Polymer encapsulated stationary phases: Advantages, properties and selectivities, Chromatographia, 27, 535, 1989.
30. A. Nahum and C. Horvath, Surface silanols in silica-bonded hydrocarbonaceous stationary phases, J. Chromatogr., 203, 53, 1981.
31. G. Schomburg, A. Deege, J. Koehler and U. Bien-Vogelsang, Immobilization of stationary liquids in reversed- and normal-phase liquid chromatography, J. Chromatogr., 282, 27, 1983.
32. G. Schomburg, J. Koehler,H. Figge, A. Deege and U. Bein-Vogelsang, Immobilization of stationary liquids of silica particles by Y-radiation, Chromatographia, 18, 265, 1984.
33. R. Haefele and D. Gjerde, Quality control and purification of oligonucleotides on the WAVE nucleic acid fragment analysis system. Application Note 103, Transgenomic, Omaha, NE.
34. C. Horvath, W. Melander, I. Molnar, and P. Molnar, Enhancement of retention by ion-pair formation in liquid chromatography with nonpolar stationary phases, Anal. Chem. 49, 2295, 1977.
35. J. C. Kraak, K. M. Jonker and J. F. K. Juber, Solvent generated ion exchange systems with anionic surfactants for rapid separation of amino acids, J. Chromatogr. 142, 671, 1977.
36. P. T. Kissinger, Comments on reverse-phase ion-pair partition chromatography, Anal. Chem., 48, 883, 1977.
37. B. A. Bidlingmeyer, S. N. Deming, W. P. Price, J. B. Sachok, and M. Petrusek, Retention mechanism for reversed phase ion-pair liquid chromatography, J. Chromatogr., 186 419, 1979.
38. A. Azarani, Patent application, Transgenomic 2000.
39. D. T. Gjerde, R. M. Haefele, and D. W. Togami, Method for performing polynucleotide separations using liquid chromatography, U. S. Pat. 6,017,457, 2000.
40. D. T. Gjerde, R. M. Haefele, and D. W. Togami, System and method for performing polynucleotide separations using liquid chromatography, U. S. Pat. 5,772,889, 1998.
41. D. T. Gjerde, R. M. Haefele, and D. W. Togami, Apparatus for performing polynucleotide separations using liquid chromatography, U. S. Pat. 6,030,527, 2000.
42. D. T. Gjerde, R. M. Haefele, and D. W. Togami, Liquid chromatography systems for performing polynucleotide separations, U. S. Pat. 5,997,742, 1999.
43. Y. Kato, Y. Yamasadi, A. Onaka, T. Kitamura, T. Hashimoto, T. Murotsu, S. Fukushige and K. Matsubara, Separation of DNA restriction fragments by high-performance ion-exchange chromatography on a non-porous ion exchanger, J. Chromatogr., 478, 264, 1989.
44. W. Bloch, Precision and accuracy of anion-exchange separation of nucleic acids, U. S. Pat. 5,856,192, 1999.

4
DHPLC

4.1
Introduction

The previous chapters have described many of the general parameters within DNA Chromatography, and how these can be controlled to achieve high resolution separations of nucleic acids. In Chapters 2 and 3, brief descriptions were given with respect to controlling the column oven temperature for the purpose of detecting mutations or polymorphism under partial denaturation conditions. Underhill and Oefner first described this approach in a 1996 publication calling the technique "Denaturing HPLC" or "DHPLC" for short [1]. Since that time DHPLC has been the most well known of all the DNA Chromatography techniques with more than 100 publications in this area alone.

DHPLC is based on the detection of heteroduplex DNA (in the presence of homoduplex DNA). A pair of primers are designed to generate a PCR product of (usually) 600 bp spanning the sequence region of interest. Individuals who are heterozygous for a single nucleotide polymorphism (SNP) have a 1:1 ratio of wild type and mutant DNA. Heating the PCR product to 95 °C and cooling slowly hybridizes the material to form a mixture of heteroduplex and and homoduplex species (Figure 4.1). In cases where the Individual possesses a mutation in both alleles, (a homozygous mutation), wild type PCR product is mixed with the sample PCR material and then the mixture is hybridized to form the heteroduplex and homoduplex species.

The heteroduplex contains a physical "bubble" of the two strands at the mutation site caused by a disruption in the hydrogen bonding at the site of the mismatch. The fundamental process governing the separation of heteroduplexes from homoduplexes in DHPLC is an increase in a heteroduplex fragment's overall single-strandedness or an increase in polarity of the fragment resulting in reduced adsorption on the separation medium's surface. A larger proportion of the hydrophilic "core" of the DNA double-helix is available for hydrogen bonding to the aqueous component of the eluent. The heteroduplex more easily desorbs from the column bead surface and is solvated by the eluent (as compared to its less single-stranded, homoduplex counterpart). At elevated temperature, this mismatch bubble becomes enlarged relative to the corresponding homoduplex. If the temperature is increased

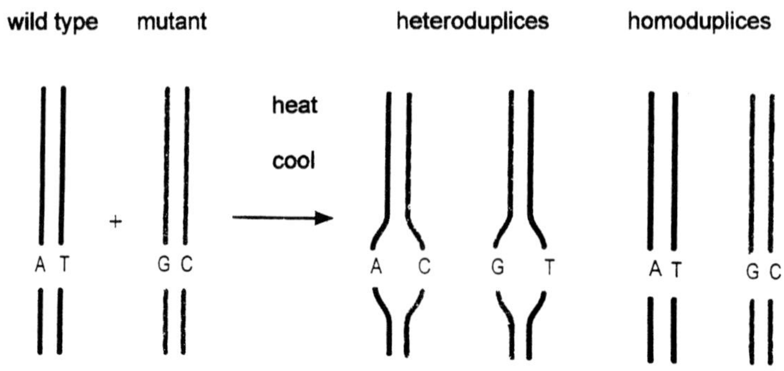

Figure 4.1. Two heteroduplex species and two homoduplex species are formed when a mutation is present.

too much, all duplex species will completely melt into single strands and will not be separated from each other. Under ideal conditions (choice of primers, column temperature, fragment characteristics, etc.), four peaks will be detected – two each for each duplex (Figure 4.2). More likely, only 2 or 3 peaks will be detected indicating the presence of a mutation (see examples throughout this chapter).

The reason for the rapid growth of DHPLC has much to do with the problem it is solving: the need to detect (as small as) a single nucleotide change in a sequence up to 1000 nucleotides long. Many techniques have been developed and studied to solve this difficult problem, each achieving varying levels of success. To describe all of them here in detail would be beyond the scope of this book, particularly in view of the excellent reviews that exist on the general topic of mutation detection [2]. For years there have been, and will continue to be, the strongest need for simple, rapid,

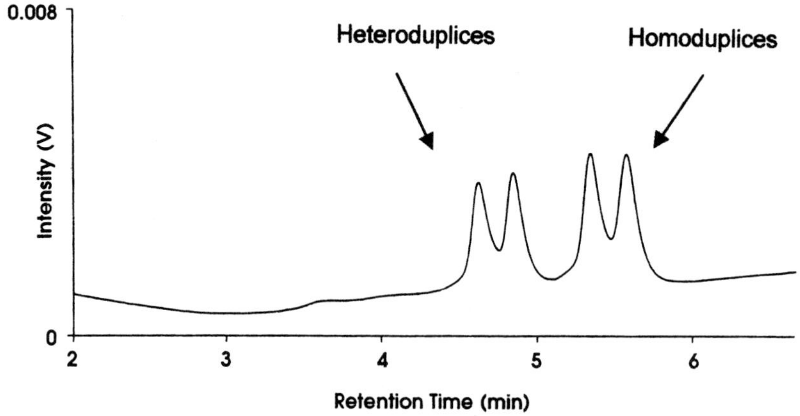

Figure 4.2. Under ideal conditions, DHPLC will produce a 4 peak pattern: 2 for each heteroduplex complex and 2 for each homoduplex. Usually heteroduplices will not be completely resolved and only a 2 or 3 peak pattern will be observed.

automated and gel-free mutation scanning techniques. DHPLC meets this need, and it is the purpose of this chapter to inform the reader as to its general principles, as well as essential points regarding its practice.

4.2 Practice of the Technique

4.2.1 Melting Phenomena and Domains

To gain a complete understanding of how to select analytical conditions to achieve these separations, one must first understand the basic phenomena involved in DNA denaturation or "melting". By understanding how DNA melts, it is possible to select the appropriate temperature with which DHPLC can be used to scan for unknown mutations within a fragment or to look for specific mutations.

DNA melting is a cooperative process. An often-invoked analogy for the description of DNA denaturation is a line of interconnected, buoyant pontoons floating on the surface of a body of water. If one of the pontoons is forced below the water's surface, then the others will also be forced below the surface as a result of their being interconnected. In other words, the buoyant bodies are "cooperatively" floating or submerged. In most cases, DNA denaturation occurs in much the same way. If the hydrogen bonding of two base pairs in the DNA double-helix is disrupted as the result of any particular process (e. g. the presence of heat, chemical denaturants, mismatched base-pairs, or combinations of any of these forces), the bases surrounding them are correspondingly disrupted and the segment of DNA denatures as a whole. In other words, the DNA denatures "cooperatively". The segment in which cooperative melting occurs is referred to as a domain, and is typically anywhere from 50–300 bases in length. Domains that are rich in G – C nucleotide hydrogen bonding melt at higher temperatures than domains that are rich in A – T hydrogen bonding. The detailed processes that govern the formation of domains, as well as the chemical processes involved in their melting has been described in detail elsewhere [3–5].

As discussed, a normal (homoduplex) fragment is composed of domains that melt cooperatively at different temperatures. The mismatch within the heteroduplex will also act in a domain manner by depressing the melting temperature of the entire domain in which it resides and not just the sequence in direct contact with the mismatch site. Under the correct analytical separation conditions, the mismatch-containing domain within the heteroduplex will be considerably more denatured than its homoduplex counterpart. As stated above, this increase in single-strandedness for the heteroduplex means that its ability to remain on the surface of the resin is significantly reduced relative to the homoduplex. In other words, the destabilizing effect of the base mismatch is effectively exaggerated or "amplified" as a result of cooperative domain melting behavior. Collectively, this gives rise to separation patterns (chromatograms) that are largely more complex (either have more peaks or at least has a much broader peak) in cases where a mutation is present, compared to when a mutation is not present.

It is possible to select the DHPLC column oven temperatures without considering the effects of cooperative domain melting behaviors. If a homoduplex fragment peak's chromatographic retention time is plotted as a function of the column oven temperature, one will often observe a well-defined sigmoidal plot similar to that shown below in Figure 4.3. It appears that this fragment may have more than one domain due to the different slopes of the inflection of the melt curve. However, any domains present are now represented by a single curve extending about 54 to 59 °C [8].

It should be noted that these plots are analogous to "perpendicular" denaturing gradient gel electrophoresis (DGGE) experiments, which are performed for the empirical determination of the optimum gradient denaturation gel conditions. Like its DGGE counterpart, this sigmoidal DHPLC plot allows for the empirical selection of optimum denaturing conditions. Using this plot, the optimum DHPLC temperature has been determined to correspond to 25 % denaturation. (25 % reduction in retention time). Another method to select the oven temperature using this plot is to take the midpoint temperature T_M and substract 1 or 2 degrees. Details of the temperature selection process are described in more detail later in this section and in Appendix 3.

The temperature is selected so that the homoduplex has just started to melt. By selecting this temperature, it is assumed that a heteroduplex, if present, will melt to a greater extent, by approximately 1–2 °C. If the midpoint is taken to be 56 °C, then subtracting 1 degree and running the sample at 55 °C seems logical. Subtracting 2 degrees would bring the oven temperature down to 54 °C and would be too

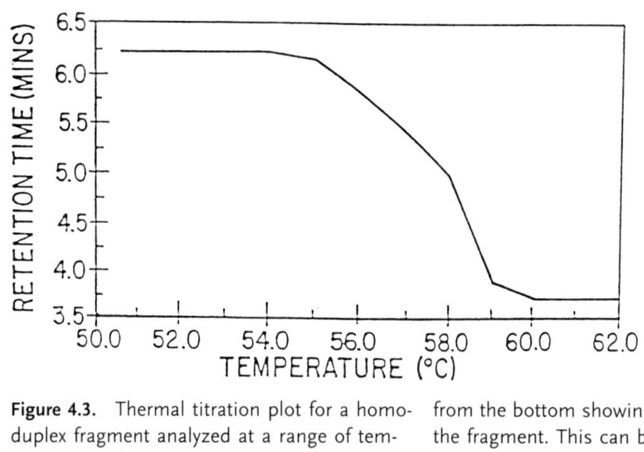

Figure 4.3. Thermal titration plot for a homoduplex fragment analyzed at a range of temperatures. When making a plot of this type it is important that the fully denatured fragment is not eluting with the void volume, but is at least partially retained on the column. If this is not done, then the plot will be (in effect) cut off from the bottom showing premature melting of the fragment. This can be avoiding by knowing the void time of the instrument (ask your instrument representative) and making sure that the faster peak elution is at least a couple tenths of a minute later than this.

low. However the plot is used, the goal is to provide the best balance-point between the non-denaturation of homoduplices (to keep their retention as long as possible) and denaturation of heteroduplices (to make them elute as early as possible).

The advantage of this empirical approach is the performance of the system used to make the temperature prediction is the same that is used to make the actual measurements. Any variation in columns, eluents, oven calibration, etc are taken into account as part of temperature selection process. The disadvantage to this empirical approach is fragment melting is usually more complex than what is shown by this plot.

There may be more than one domain present. If the homoduplex melting plots are steep and the temperature range less than 5 degrees, then it is usually possible to screen for mutations in both domains by selecting a single (just starting to melt) analysis temperature. However, the temperature selected must be less than or equal to the melting temperature of the domain that has the lower melting temperature. If an intermediate temperature is selected, the lower domain in both the heteroduplex and homoduplex fragments may be already denatured and the ability to detect mutations in that domain will be lost.

This multi domain issue becomes more severe when the melting range is greater than 5 °C. More than 1 temperature will probably have to be used to ensure that the whole fragment is screened for mutations. For example, if a DNA fragment contains three domains that melt at 55, 58 and 64 °C, the slope of the melting plot will extend over 10 °C range and will be broad. The lower two melting domains can probably be simultaneously screened at a temperature of 54 °C and the higher melting domain screened at 62 °C. However, there is no way to know if these temperatures are the best choices since the exact melting profile of the fragment is not known.

In most cases the fragment does not have a single, uniform melting domain. This multi-domain characteristic can often render the temperature melt plot less reliable. These problems can be largely overcome through ab initio prediction and plotting of a fragment melt map and then interpretation of fragment melting behavior. How these melt maps are generated and how they are interpreted for DHPLC operating temperatures are the topics of the next section.

4.2.2 Temperature Prediction

The ability to predict a fragment's melting behavior *ab initio* has more than one benefit associated with it. For one, it is possible to predict a single set of analytical conditions for mutation detection. In a highly ambitious study that involved DHPLC mutation analysis of many hundreds of different fragment sequences, investigators determined that a single DHPLC analysis temperature provided 87 % mutation detection sensitivity, with no attention being given to multi-domain melting behaviours (or primer design) whatsoever [6].

The second benefit associated with melting prediction is that it provides the means necessary for obtaining virtually 100 % mutation detection sensitivity. As noted above, a multi-domain fragment can result in less reliable determination of temperature, since more than one melting domain can result in a bias to the predicted "optimum temperature". However, with *a priori* knowledge of more than one melting domain, it is possible to avoid improper selection of the analysis temperature.

It is helpful to illustrate the above points by way of example. Shown in Figure 4.4 are two plots for a particular sequence of 300 bp long. The first plot shows "Helical fraction (%)" as a function of "Temperature". The second plot shows the melting "Temperature" as a function of base position, and is well known to practitioners of DGGE as a "melt map".

There are some observations that need to be made with respect to these plots. The first observation is that the first plot ("Helical fraction vs. Temperature") is largely the same as that obtained when one plots the empirical data of "Peak retention time" versus "Analysis temperature". The second observation is that the overall shape of the first plot is the result of the data shown in the second plot (Temperature vs. Base-pair). A third observation is that there are different domains within the second plot. These domains are characterized by stretches of bases that possess largely identical melt temperatures (e. g. Domain 1 between bases 107 and 300; the "fraying" of the bases at both ends can be largely disregarded, since this will represent sequences covered by the PCR primers). It is within these domains that nucleotides are subjected to the "cooperative" melting processes described earlier. Therefore, each domain can sometimes (but not al-

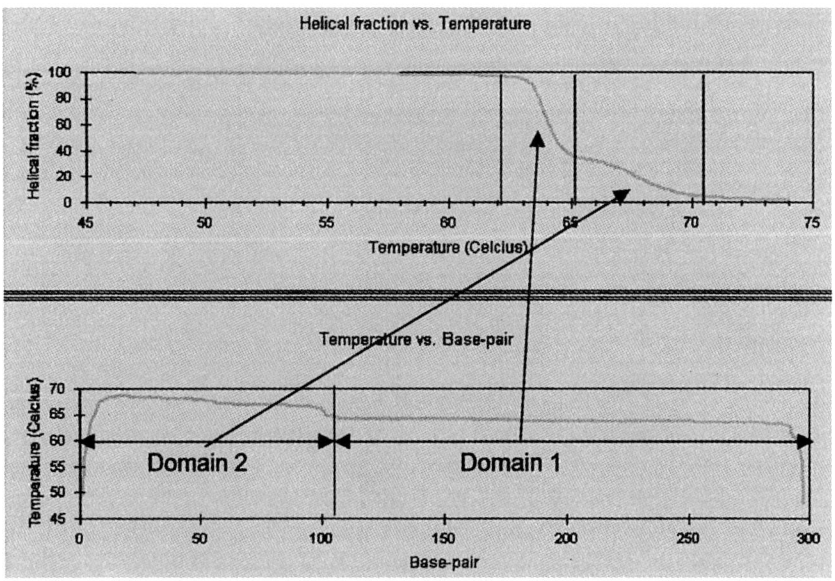

Figure 4.4. Correlation of predicted thermal titration data (shown at the top of figure) and melt map features (shown at the bottom of the figure). Predictions performed with WAVEMaker® software.

ways) require its own analysis temperature for successful mutation detection within that domain. For the example shown below, domain 1 would require an analysis temperature ~63 °C (i. e. roughly 25 % denaturation of domain 1 alone), and domain 2 would require an analysis temperature of ~67 °C (i. e. roughly 25 % denaturation of domain 2 alone).

This is not to say that more than one temperature must always be used for a fragment that has more than one domain. Skopek and co-workers [7] performed an excellent study whereby exon 8 of HPRT was point-mutagenized such that 20 separate mutations were introduced across the entire sequence. Twelve mutations were introduced in the low-melting mutant domain and 8 in the high-melting mutant domain. There was a 5 °C difference in melting temperature between the lowest-melting and highest-melting mutant. They found that all mutants were detected at a single column temperature but that detection was clearly at the expense of resolution between heteroduplex and homoduplex species (and therefore, confidence in the final result). It bears noting that the high-melting mutant domain was adjacent to an even higher-melting domain that possessed no mutants. The highest domain was perhaps acting as a clamp (discussed later) to keep the fragment together and facilitate the detection of mutations in lower melting domains.

There are a variety of software packages available that can predict melt maps of this sort. To perform these predictions, one must have information regarding the hydrogen bonding energies of nucleic acids; these energies are reported in the literature [5]. In longer DNA fragments (i. e. PCR products), an intrinsic helical tendency is combined with a conditional probability such that the probability of a base being in the helical state is strongly affected by its neighbors; this is what gives rise to cooperative melting, and the domain behaviors described. Virtually all of these predictive algorithms are recursive in nature, such as the implementation by Fixman and Friere [4] of Poland's original model [3].

These algorithms have been optimized to predict melting behavior in solution states different from those commonly applied in DHPLC. For the purposes of DHPLC these predictive algorithms must be adapted to reflect the presence of acetonitrile (an organic solvent) and triethylammonium acetate (a salt and ion pairing reagent) in the separation environment (a nonpolar surface), since both of these components have significant effects on fragment melting behavior. An organic solvent will decrease the fragment melting temperature, a salt will increase temperature, and in the presence of a ion pairing reagent, a nonpolar surface will increase the temperature. The extent to which these components influence fragment melting have been described in detail [8]. DHPLC algorithms of this nature are available commercially through WAVEMaker® software (Transgenomic, Inc.) [9, 10] or via the Internet (http://insertion.stanford.edu/melt.html). Once these operations are performed, it is possible to accurately predict the analytical conditions required for complete DHPLC analyses: the analytical separation gradients are generated, and the analytical temperature is predicted in the manner described above. This general predictive approach to DHPLC analyses has been validated on numerous occasions (see references in the "Review of DHPLC Publications" section). In summary, it has been shown that this approach can detect anywhere from 87 % of mutations (single temperature selection with no fragment or PCR optimization whatsoever) to 100 % of mutations (one or more temperatures selected with fragment optimization).

4.2.3 Primer Optimization and Clamping

Temperature prediction alone will not always provide conditions that are fully optimized for the detection of every possible mutation. This is particularly the case when attempting to detect mutations located in high-melting domains within a fragment. This situation is addressed through PCR primer optimization. In the context of DHPLC, what is meant by PCR primer optimization is the selection of primer sequences that give rise to fragments whose melting characteristics are propitious with respect to the detection of mutations within a particular sequence. Of course any primers that are optimized for DHPLC must also be appropriate for PCR amplification.

These points are well illustrated in the example shown in Figure 4.5 [8]. These are melt maps for the same mutation, but with different primers selected. It is clear from Trace #1 that the mutation resides in a low-melting domain, and that the vast majority of the fragment (roughly 75 %) is unmelted at that temperature that corresponds to the domain that possesses the mutation. In Trace #2, the mutation also resides in a low-melting domain, but a smaller proportion of the fragment (roughly 40 %) is unmelted at that temperature that corresponds to the domain that possesses the mutation. In Trace #3, this proportion is reduced even further (roughly 15 %). Important to note here is that the fragment size is identical in each case, and all that varies is the proportion of unmelted fragment at the mutation domain's melting temperature.

Figure 4.6 shows the consequence of these differences [8]. The separation shown in Figure 4.6a is not only sufficient for baseline resolution of the heteroduplices from the homoduplices, but also for the separation of these duplices from one another.

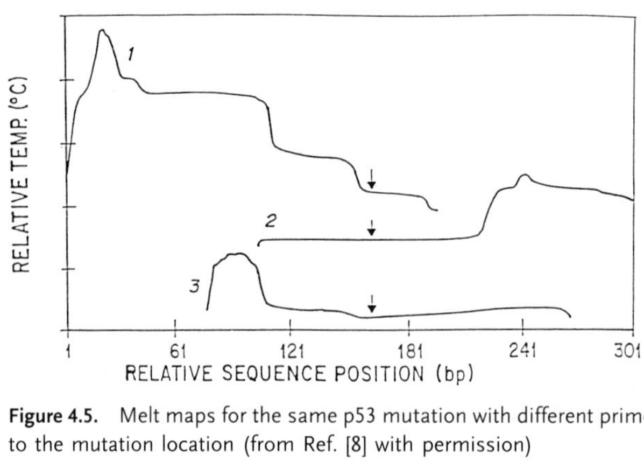

Figure 4.5. Melt maps for the same p53 mutation with different primer locations selected relative to the mutation location (from Ref. [8] with permission)

In Figure 4.6b the resolution is clearly diminished, but still allows for highly reliable detection of the mutation. In Figure 4.6c the heteroduplex is barely resolved from the homoduplex. There still is enough resolution for detection of the mutation, especially when using computer analysis, the detection is clearly more difficult.

In all three cases the mutation is situated in a low-melting domain, which is why it is successfully detected in all three cases. However, our degree of analytical con-

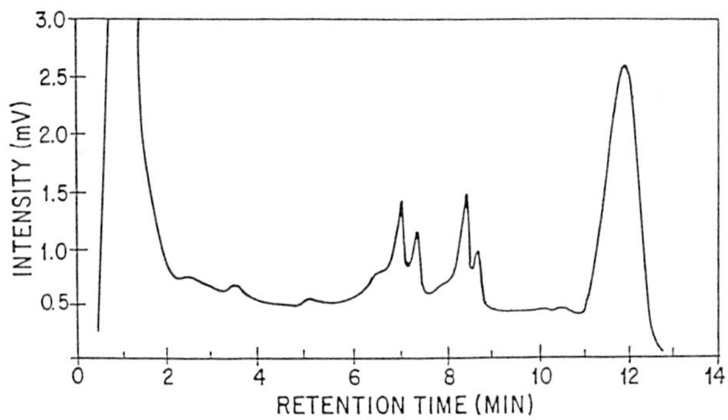

Figure 4.6a. DHPLC chromatogram of fragment 1 of Figure 4.5 (from Ref. [8] with permission).

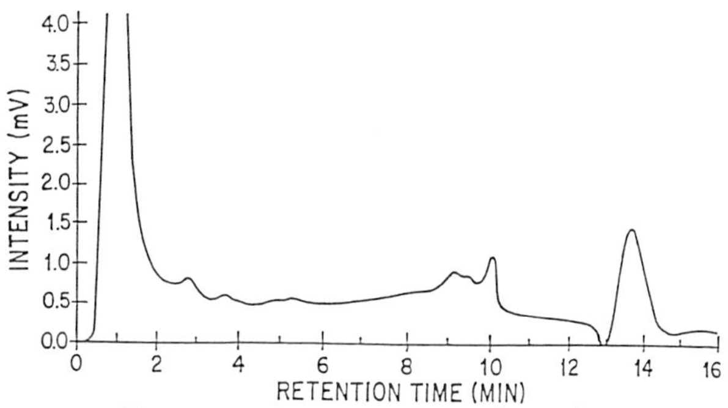

Figure 4.6b. DHPLC chromatogram of fragment 2 of Figure 4.5 (from Ref. [8] with permission).

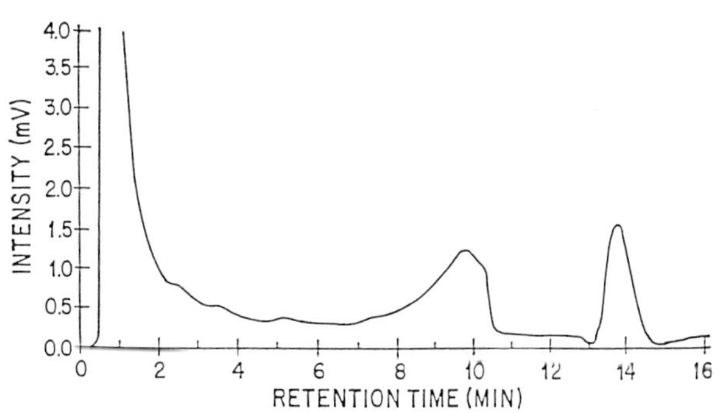

Figure 4.6c. DHPLC chromatogram of fragment 3 of Figure 4.5 (from Ref. [8] with permission).

fidence is always greatest when the resolution for any given mutation is at its maximum. Therefore, maximization of detection confidence is achieved through proper primer optimization strategies that situate a sequence region for scanning in a low-melting domain with a high proportion of unmelted sequence adjacent to it, which is demonstrated here and elsewhere [7]. The needs for the level of confidence must always be balanced against the detection productivity of the procedure. In other words, maintaining the absolute highest level of confidence at all times may necessitate analyzing numerous overlapping fragments for the presence of mutations, whereas fewer fragments can be scanned when the confidence level need not be at the absolute highest levels.

Even in cases where the confidence level can be relaxed, there are still circumstances where primer optimization does not yield conditions that allow reliable mutation detection. This is frequently the case when the region being scanned is in a high-melting domain, and no amount of primer optimization can render this domain more amenable to mutation detection (or, the scanned region is in a high-melting domain and it is not possible to move the primer locations). A common means of overcoming this situation is the addition of a non-template GC-clamp to one of the primers used for amplification, which has been shown to work successfully in DHPLC [11]. By doing so, one creates an additional high-melting domain adjacent to the region being analyzed for mutations. This general approach has been applied for over a decade in the context of DGGE, and is largely compulsory for the successful application of that technique.

However, in DHPLC GC-clamps need only be applied in extreme circumstances such as those described above. An example of this approach is shown in Figure 4.7 for mutation detection within exon 1 of k-*ras*, where the first melt map shows that

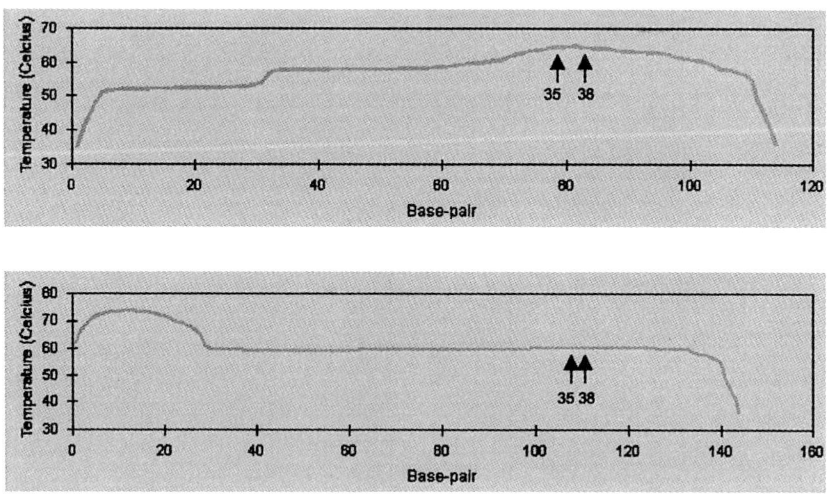

Figure 4.7. DNA melt maps for the amplified exon 1 ras sequence. The first melt map shows the melting behavior without any modification of the PCR primers, and the second melt maps shows the melting behavior with a 30-bp GC-clamp attached to the forward primer.

the region requiring mutation scanning (i. e. the region that contains genomic sequence base positions 35 and 38) is in the highest melting domain. In this instance it was not possible to move the primer locations, so a 30 bp GC-clamp was added to the forward PCR primer. This gave rise to the second melt map, which shows that the mutation is in a low-melting domain since the GC-clamp confers a high-melting domain adjacent to it. This example will be returned to later in this chapter.

4.2.4
PCR Fidelity

The detection of mutations by DHPLC requires the chromatographic separation of heteroduplices from homoduplices. The mutant sequence present in the template is carried through the PCR amplification process, the heteroduplices are subsequently formed through standard hybridization protocols, and the separation is commenced. While this is a generally straightforward procedure, one must always be on guard for the potential introduction of confounding artifacts brought about by the analytical procedure itself. One part of the DHPLC analytical technique where artifacts may be introduced is within the PCR process itself. PCR induced mutations or amplification errors of sample can make detection of the sample mutations difficult or even impossible.

It is well known that DNA polymerase will misincorporate non-complementary nucleotides in the process of template duplication. While the rate of these PCR induced mutations is seemingly small (i. e. 1 per every 50,000 nucleotides $= 2 \times 10^{-5}$), these introduced errors are exponentially amplified along with the faithful duplication of the sample nucleotides. These imperfections of the PCR process usually do not present any significant limitations to most DHPLC analyses as long as certain precautions are made. These include using preparing the sample and standard together for a particular experiment so that if there are errors, then they are the same for sample and standards. a proofreading polymerase can be used as well. As the need for detecting low concentrations of heteroduplices increases, the ability of PCR induced mutations (and their associated heteroduplices) to confound the final detection of the mutation increases.

Differences in PCR fidelity, and how they present themselves in the DHPLC signal profiles, are shown in Figure 4.8 [8]. For a given set of amplification conditions with a homozygous template, it is clear that the non-proofreading AmpliTaq® polymerase results in a significantly higher relative amount of background heteroduplices, whereas the proofreading *Pfu*® and *Pfu*Turbo® provide significantly reduced levels of this background [8]. It is clear from these examples that the AmpliTaq®-derived signal could lead to misclassification of the sample being analyzed, whereas this is less likely for the higher fidelity results. Properly optimized amplifications can minimize the risk of misclassification with non-proofreading enzymes, while proofreading enzymes will always minimize the risk posed by the presence of background heteroduplices. The need for the lowest possible levels of background heteroduplices is most pronounced when attempting to detect mu-

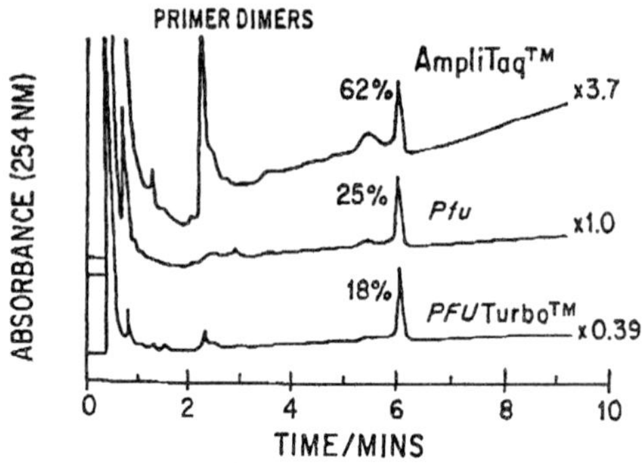

Figure 4.8. Background heteroduplex signals generated as a result of PCR infidelities. The background signal is reduced upon application of a high-fidelity PCR polymerase.

tant sequences at concentrations significantly lower than the 50 % found in germline mutation detection scenarios. These types of determinations are the subject of the next section.

4.2.5
High-sensitivity DHPLC Determinations

Recently there has been a great deal of interest in the ability to detect and accurately quantify low concentrations of mutant alleles in the presence of a high concentration of normal/wild type alleles. Some examples of recent work with other methods include constant denaturant capillary electrophoresis (CDCE) [12] and DGGE [13]. A compelling aspect of DHPLC is that its separation ability provides a natural means to perform both quantitative analysis and/or purification of mutant sequences.

Assuming that the mutation is heterozygous, the amount of a germline mutant sequence is effectively identical to the amount of normal sequence rendering quantitative capabilities largely irrelevant. On the other hand, there are numerous instances where the ability to accurately quantify and potentially purify mutant sequences in the presence of a high amount of normal sequence would be highly desirable. Examples of this include pooling of samples for enhanced scanning throughput, the detection of somatic mutations in the context of cancer diagnostics, and the determination of heteroplasmy (low levels of mutated cells in the presence of normal cells). DHPLC-based analyses for these types of determinations have been described in the literature, and will be discussed in greater detail in the "Review of DHPLC Publications" section below. While these studies have demonstrated the general feasibility of using DHPLC for the detection of low mutation concentrations, there is little information as to how various analytical parameters influence the determination and introduce fundamental limitations. The next several sections will discuss these limitations. Section 4.2.5.4 will use an example application, the detection of low concentrations of particular k-ras alleles, to examine the factors involved in high-sensitivity DHPLC, along with strategies for optimization of these factors.

4.2.5.1 Chromatographic Resolution between Heteroduplices and Homoduplices

The detectability of a particular component is directly related to the extent to which it is resolved from other potentially overlapping (i. e. "contaminating") components. The term resolution is often cited when attempting to calculate the degree of separation between two peaks. The resolution between peaks, or R, is calculated as:

$$R = \frac{2\,\Delta\,t}{(w_{p1} + w_{p2})}$$

where Δt is the difference in retention times between the two peaks in question, and w_{p1} and w_{p2} are the peak widths (in time units) for peak 1 and peak 2, respectively ($w_p = 1.70 \times w_{1/2}$). This gives a unitless quantity that describes the relative extent of component retention (i. e. the numerator), as well as the efficiency to which they are separated (i. e. the denominator). If it is assumed that the mutation-harboring heteroduplex peak is our target for detection, it is clear in the case of complete peak overlap (R=0) that detection is impossible since it is entirely "contaminated" by the coeluting homoduplex peak. As the target peak becomes more resolved (i. e. increasing values of R), accurate detection and quantification become more likely. As we increase R, we decrease the extent to which we are forced to measure the heteroduplex peak on top of a shifting background signal.

It is certainly possible to detect a germline mutation by poorly resolved single peak in front of the homoduplex peak (a very small increase in R) even a simply broadening of the broadening homoduplex peak (see normalization of peak discussion in Chapter 2). To maximize the confidence in the measurement of minority alleles, it is necessary to maximize the degree of resolution the R to 1 or more. To do so, one must design DNA fragments whose melting characteristics provide the greatest degree of resolution. In general terms, this is achieved by ensuring that the region(s) of DNA being scanned for mutations reside in low-melting domains within the fragment. Also, as indicated in Section 4.1.3, a relatively high proportion of unmelted sequence adjacent to the low-melting sequence will further enhance this resolution.

4.2.5.2 Mass Sensitivity for the Resolved Heteroduplices

Assuming that we now have adequate resolution the next fact is to ensure that the detector is capable of detecting the necessary mass of the resolved minority component. If adequate mass sensitivity is not achievable, then the effort involved in obtaining suitable resolution of the components is wasted.

A standard means of detection applied in DHPLC is that of UV absorbance at 260 nm. This mode of detection provides mass detection limits on the order of 0.3 ng of DNA which is comparable to what can be obtained with a standard ethidium bromide gel analyses. While this certainly is a very respectable limit of detection, it bears closer scrutiny. For example, if one wishes to detect a mutant allele population of 1 %, then the total amount of DNA fragment that must be introduced to the system is 30 ng. However, if one wishes to *quantify* at the 1 % level with a high degree of precision, then the amount of material introduced to the system would need to be anywhere from 5–10 times greater (150–300 ng). While the majority of DHPLC systems that are described in the open literature can analyze this amount of material, the quantities approach the practical analytical column capacity of 500 ng. One can simply accept having broader peaks as the column capacity is being reached or one may use a high capacity, preparative column.

As well as approaching the column capacity limit, there can also be other factors that contribute to lower detector signal intensities. For example, the amplification may be performed from archival samples (i. e. paraffin-embedded tissues). In this instance, one may wish to amplify only a short stretch of sequence so as to minimize the invariable effects of genomic DNA degradation in archival samples. Since the PCR-generated fragment may be significantly shorter than what might be typically encountered (say <100 bp from archival samples versus 500 bp from fresh samples), its overall molar absorptivity is significantly lower. This is due to the fact that a given molarity of a 100 bp DNA fragment possesses five-fold fewer chromophores than the same molarity of a 500 bp DNA fragment, thus reducing its overall detectability.

Furthermore, the conditions surrounding the PCR amplification itself may contribute to a decrease in the overall signal intensity. For example, in the interest of minimizing the effects encountered as a result of PCR-induced mutations it is common practice to minimize the number of PCR cycles. While this reduces the aforementioned "PCR noise", it will naturally lead to a reduction in the overall PCR product yield. This would also serve to reduce any given fragment's overall detectability, regardless of size or sequence content. All of these factors can contribute separately and/or simultaneously, depending upon the needs for the particular determination.

A common means for providing additional mass sensitivity in DHPLC is through the use of fluorescent labels, which has been described elsewhere [14]. The primary benefit engendered by the use of fluorescent tags is that they provide a significantly enhanced signal-to-noise ratio over UV absorbance for an equal quantity of DNA. While this enhanced sensitivity can lead to greater precision within the measurements, it does not in and of itself provide lower detection limits for a highly resolved heteroduplex fraction. Detection limits in general are defined by the International Union of Pure and Applied Chemistry (IUPAC) as the concentration that gives rise to a signal that is equivalent to three times the standard deviation of the blank ("zero analyte") signal. While having enhanced sensitivity for DNA could certainly contribute to the reduction of the blank's standard deviation and hence detection limits, another factor which will contribute even more so is that of spurious PCR-induced "background mutations" being introduced into the measurements. If this is the primary source of the blank signal, it can introduce a major fundamental limitation to any minority allele determination. It is the topic of PCR-induced "background mutations", or "PCR noise", to which we now turn our attention.

4.2.5.3 **PCR-induced Background**

As noted above, spurious mutations can be introduced to the sequence under analysis by the PCR process itself. These PCR-induced mutations are the result of what is commonly referred to as "PCR infidelity", which is a well-known characteristic of PCR in general. The issue of PCR fidelity has been studied in depth [15], a full treatment of which is beyond the scope of this paper. Rather, we will present how the issues of PCR infidelity present themselves in the context of high-sensitivity DHPLC determinations.

It has been shown [15] that DNA polymerases used for in vitro DNA amplification do not replicate the nucleotide sequence in a perfectly faithful manner. When a polymerase attempts to incorporate a complementary base, it can occasionally make errors in doing so. The polymerase error rate, or p, is the number of errors committed per nucleotide. Therefore, if a polymerase has an error rate $p = 2 \times 10^{-5}$, then the polymerase misincorporates one base for every 50,000 bases replicated (since $1/50{,}000 = 2 \times 10^{-5}$). While this may seem like a vanishingly small error to merit any attention, a number of issues must be remembered. Firstly, each PCR-induced error that occurs will subsequently be replicated and amplified in the subsequent PCR cycles, and any new errors that are added will subsequently be replicated and amplified, so on and so forth. When given a polymerase error rate p, it is possible to approximate the percentage of fragments that carry PCR-induced mutations [16]. The error frequency per nucleotide f, is given as:

$$f = np/2,$$

where n is the number of PCR cycles. Therefore, if the number of cycles $n = 30$ and the polymerase error rate $p = 2 \times 10^{-5}$, then the error frequency per nucleotide $= 3 \times 10^{-4}$ (or, 1 in every 3,333 bases carries a PCR-induced mutation). However, since this is a *per nucleotide* error frequency, it can result in a significant percentage of fragments carrying a PCR induced mutation. If it is assumed that 500 base-pairs of DNA within a fragment are being replicated by the polymerase, then the percentage of fragments carrying PCR-induced mutations would be $(1{,}000/3{,}333) \times 100 = 30\%$!

Secondly, any and all mutation-derived mismatches within the PCR products will give rise to heteroduplices, whether the mutations originate from the genomic DNA sequence or are introduced in the PCR. The latter instance will give rise to a significant "mutant background" signal, and can lead to an overestimation of the amount of mutant present if not taken into consideration. Using the example above, the amount of "mutant background" present would be 30 %, in the presence of which we would be attempting to detect our target mutant signal. Therefore, with respect to the minimum quantity of mutant detectable by DHPLC and adhering to the IUPAC definition of detection limits, it is the variation of *the background signal itself that defines the mutant detection limit* (as opposed to the variation associated with the instrument noise alone). Therefore, the total background variation must be measured directly if a true detection limit is being determined.

It bears noting that no single detection limit can be cited for high-sensitivity DHPLC analyses, as different amplifications (and other experimental conditions) will contribute differently to the final analysis. Furthermore, it is not meaningful to provide a detection limit for a particular high-sensitivity DHPLC analysis without first having characterized the standard deviation ("noise") contributed by any and all background signals, PCR-derived or otherwise.

Therefore, each analysis must be optimized in an effort to meet the final analytical needs. For example, if the limit of detection is the ultimate priority for the analysis, then significant emphasis must be placed upon achieving a high degree of amplification fidelity (not to mention the factors of resolution and mass sensitivity noted earlier). This means applying high-fidelity polymerases and/or adopting chemical amplification conditions that lend themselves to enhanced PCR fidelity [16], such as fully optimized temperatures, pH, [Mg^{2+}], or [dNTPs]. Conversely, if a modest pooling factor is being applied for screening experiments and the requirements regarding PCR-induced "background" are somewhat relaxed, then an enzyme with no proofreading capability (and hence significantly reduced cost) can be applied.

4.2.5.4 Example Application: Detection of Varying Levels of k-ras Alleles

In the example discussed in this section, different levels of two separate k-ras alleles are determined in the presence of wildtype k-ras sequence. The mutations occurred within codon 12 at genomic nucleotide position 35 (GGT→GAT), and within codon 13 at genomic nucleotide position 38 (GGC→GAC). These coding-sequence mutations give rise to the same amino acid change, namely G12D and G13D. When the sequence was originally amplified with the template-sequence primers, this gave rise to the first melt map shown earlier in Figure 4.7. Note that nucleotide position 35 and 38 in the genomic sequence now correspond to base numbers 79 and 82 indicated in the figure, respectively, since the amplification product also encompassed the intronic sequence 5′ to exon 1. From the examination of this melt map, it is clear that the mutations under examination reside in a location that is not propitious for high-resolution separation. If the target mutations had resided in the amplified fragment's first ~60 bases (corresponding to the intronic sequence plus the first five codons), there would be a reasonable chance to detect the mutations in the fragment, since these bases are situated in low-melting domains adjacent to a relatively high-melting domain. As it is in this fragment, the target mutations (shown by the arrows) actually reside in the worst possible location (that is, within the absolute highest melting domain within the fragment).

To address this sub-optimal melting behavior, and since the primers could not be shifted in any way with respect to template location, a non-template high-melting domain was introduced into the fragment. This was achieved through the attachment of a 30-nt GC-clamp to the forward primer, which resulted in the second melt map shown in Figure 4.7. Also, a 6-FAM (6-carboxyfluorescein) fluorescent tag was placed on the GC-clamped forward primer so that maximum column performance is maintained while simultaneously providing a high degree of mass sensitivity. For this particular application, a proofreading enzyme is not necessary for allele detectability and determination.

The results of these analyses are shown in Figure 4.9. The first two traces show the G12D allele in the presence of three replicate analyses of the wildtype sequence. In each case the concentration of the allele was one part in 30 (3.33 %). The last two traces show a dilution series for each allele, where the series includes 0 % mutant, 3.33 % (1:30) mutant, 10 % (1:10) mutant and 40 % (1:2.5) mutant. When these signals are temporally aligned on the homoduplex peak, and the signal intensities are normalized to the homoduplex peak, then the signal remaining to the left of the homoduplex peak is directly proportional to the amount of heteroduplex present. These signals show that there is clearly a correlation between the different levels of mutant and the normalized heteroduplex signal intensity. Resulting calibration curves for each allele are shown in Figure 4.10, both of which demonstrate excellent linearity.

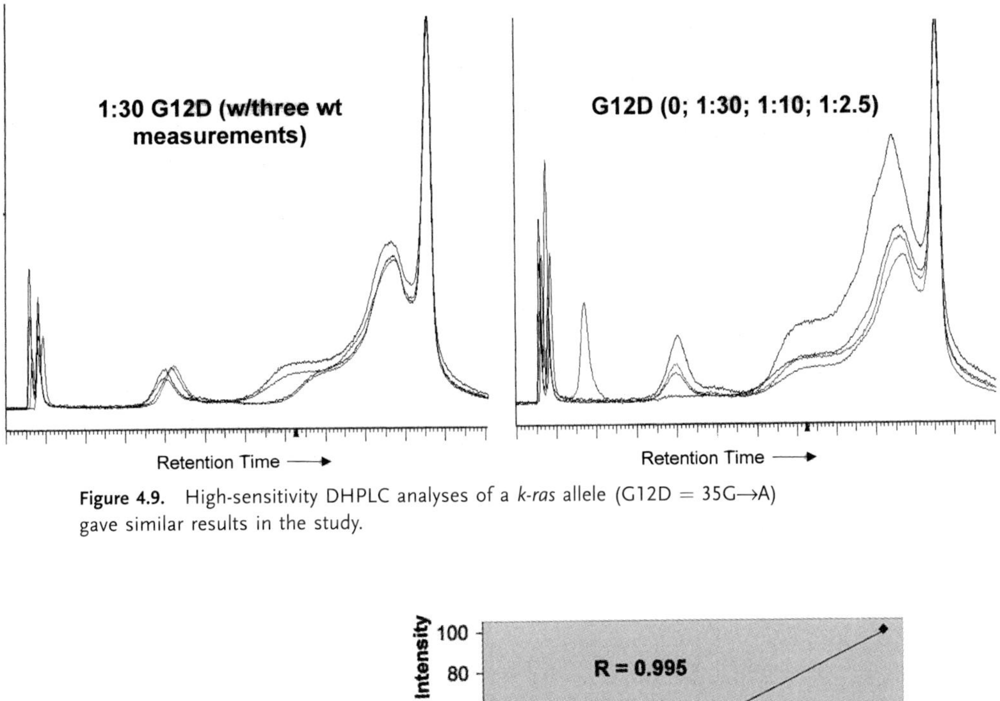

Figure 4.9. High-sensitivity DHPLC analyses of a *k-ras* allele (G12D = 35G→A) gave similar results in the study.

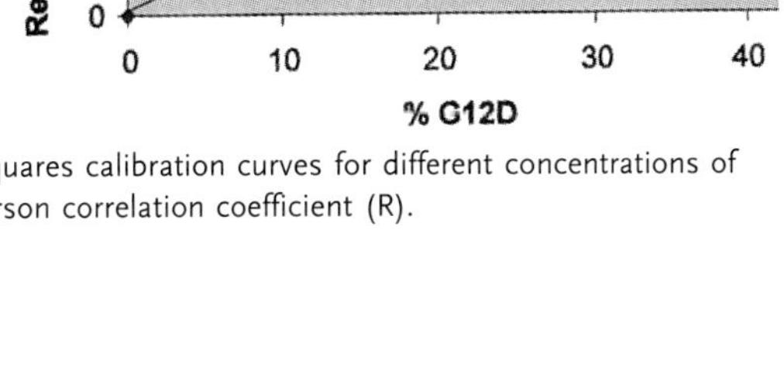

Figure 4.10. Data and linear least-squares calibration curves for different concentrations of pooled G12D allele, shown with Pearson correlation coefficient (R).

4.3
Review of DHPLC Publications

As mentioned earlier in this chapter, the overwhelming majority of DNA Chromatography publications have been in the broad application area of DHPLC. Because of the clear popularity of this approach to heteroduplex analyses, a superb and comprehensive review of DHPLC's application has been coauthored by one of DHPLC's inventors [17]. So to avoid redundancy, and to more specifically address the stated goal of this chapter to inform the reader on DHPLC's general principles and practice, only those publications that specifically address DHPLC practices will be covered here, i. e. those publications that increase our understanding of how DHPLC can or should be performed in general. Lack of a particular publication's inclusion in this review should not be taken as the authors' perception of its lack of importance. Many DHPLC publications report the final results of DHPLC analyses in a larger context, or will describe the conditions developed for a very specific analysis (such as the primer sequences and analytical conditions applied for mutation analysis of a particular gene). In other words, the goal for these types of publications is not to extend the general level of understanding regarding DHPLC phenomena as a whole, so they are not included in this review.

As mentioned earlier, the first peer-reviewed publications that designated DHPLC were in 1996 [1, 18], with a follow-up publication to the original report regarding Y-chromosome genetics in 1997 [19]. Another paper in that same year deals with DHPLC-based analysis of the candidate tumor suppressor gene PTEN/MMAC1 in glioblastomas [18]. This paper is particularly notable for a number of reasons. For one, this is the first DHPLC paper to provide specific information regarding validation of the technique, which was based on confirmatory sequencing of exons 1 and 5 for all 73 tumor samples studied. Sequencing did not reveal any mutations that were not already detected by DHPLC, nor were any "false positives" attributed to DHPLC analyses. Second, DHPLC as a technique was extended to the detection of somatic mutations in tumor samples. The authors state that they found DHPLC to be very sensitive even when the ratios of mutant-to-wildtype were very low. However, there is no indication in the paper as to exactly how sensitive it is in terms of these ratios. This lack of quantitative information was explicitly addressed the following year in a paper by the same group [21]. The authors show it is possible to distinguish as little as 10 % mutant DNA present as a heteroduplex species, where PTEN/MMAC1 was again the target gene in various glioblastomas (37 out of 40, or 92.5 %, were properly classified by DHPLC). This paper also extends the general validation of DHPLC by analyzing a series of previously sequenced CFTR and RET germline mutants. Twenty out of 22 (91 %) of these mutants were properly classified by DHPLC. This work was extended to the examination of PTEN/MMAC1 in small cell and non-small cell lung cancers [22]. The application of DHPLC in somatic mutation detection mode allowed the authors to conclude that PTEN/MMAC1 mutations have a role in small cell, but not non-small cell, lung cancers, and that p73 mutations do not have a role in lung tumorgenesis [23].

All of these papers illustrated the ability of DHPLC to provide genetic variation data in a rapid, sensitive, and quantitative manner for any genetic sequence. However, very little detail was given in any of these papers as to how analytical conditions (particularly the crucial analysis temperatures) were selected. For example, analytical temperature selection was described somewhat subjectively in more than one publication as being that temperature which gave rise to "a significant decrease in retention" [18–20] for the injected PCR products. In one of these papers [19], the authors also point out (though no detailed data were provided) that mutation detection sensitivity was improved further by using 7-deaza-2′-dGTP (in lieu of dGTP) for the PCR amplification of certain fragments. This nucleotide minimizes the stabilizing effects of GC-rich regions whose melting temperatures were 10 °C more than the amplicon's average melting temperature. Despite this lack of elaboration or supporting data, these broad observations point to the fact that melting behaviors across the amplicon (and control over them, when possible) are critical for successful DHPLC analyses.

In the following year of 1998, another paper whose authors include those from the original 1996 DHPLC publications described blind analyses of exon H of Factor IX and exon 16 of NF1 [24]. Along with 100 % concordance with sequencing results, the authors also pointed out that DHPLC signal shapes often change for different mutations. They indicated that shapes tend to be similar over significant stretches of the fragment (which is likely to be from the individual mutations being present in the same melting domain). These authors reiterate their previous claims that a single temperature can be used for fragments whose melting domains differ by as much as 10 °C, and that those with a larger temperature range should use two temperatures (despite prior recommendations of using 7-deaza-2′-dGTP and a single temperature for ranges this large [19]). No stipulation was given as to how these two temperatures should be selected.

The issue of melting behaviors across the entire PCR amplicon, as well as other analytical parameters related to DHPLC, were alluded to in a paper where investigators used a single Y-chromosome SNP as a model to describe the importance of various DHPLC parameters [25]. These parameters included PCR primer design and their respective locations on the template, the use of high-fidelity DNA polymerases (so as to minimize potentially confounding PCR-induced mutations), chromatographic gradient optimization, and proper selection of analysis temperature. The discussion of temperature selection resulted in more specific recommendations. These recommendations were based on complete on-column "thermal titration" of the entire amplimer ($T_{M,50\%\ helical} - 1\ °C$). This recommendation was in opposition to suggesting subjective judgements of a "significant decrease in retention", which can easily give an incomplete picture of amplimer melting due to the lack of knowledge regarding a fragment's remaining (i. e. unmelted) domains.

Later in 1998, the investigators who published the first three papers on DHPLC co-authored a publication that extended these melting principles even further [26]. An algorithm was introduced for predicting DHPLC analysis temperatures from computer-modeled DNA melting of user-specified DNA sequences. The algorithm is available via the Internet, and sequences were submitted for temperature predictions. While no details were given in the paper as to the algorithm's logic or scientific basis, the authors claim that triethylammonium pairing ion in the eluent had the effect of stabilizing AT base pairs, though this was not experimentally demonstrated to the conditions applied. This stabilization purportedly allowed for the use of one analysis temperature per fragment, even when melting domains differ by as much as 10 °C. In cases where melting domains differed by more than 10 °C, it was suggested that two different melting temperatures be used (though the reader is not told how the two temperatures were selected). Predicted temperatures were used along with empirically derived analytical temperatures, though no performance comparisons were made between the predicted and empirically derived temperatures.

In 1999 there were a number of papers that attempted to elucidate the precise role of various analytical parameters within DHPLC. One paper examined issues surrounding the selection of optimal analysis temperatures, and compared their results to SSCP (single-stranded conformational polymorphism method) and heteroduplex analysis [27]. Different amplimers from CFTR (cystic fibrosis transmembrane conductance regulator gene), TSC1 and TSC2 genes were used for the evaluation of a predictive algorithm, which was described in the paper and based on the original cooperative DNA melting behavior described by Poland [3]. Based on this predictive model, the authors calculated the "% helical" versus temperature. With this, the authors selected a recommended oven temperature (RT_m) corresponding to the highest site melting temperature T_i. If the fragment has a melting range of >5 °C, then a temperature 5 °C lower is also recommended. The authors analyzed 32 different heterozygous samples, and found that 31 out of 32 were detected at RT_m and at RT_m+1 temperatures. They also studied the effect of changing the analysis RT_m, and showed that sensitivity dropped off precipitously for temperatures >1 °C lower or >2 °C higher than the predicted RT_m. A plot of their data is shown in Figure 4.11, which clearly demonstrates the sensitivity of DHPLC analyses to proper temperature selection and control. The authors point out that 32 out of 32 heterozygous samples were correctly classified when the analysis performed at were RT_m and RT_m+2 °C temperatures, with the authors broadly recommending these temperatures (though separate analyses of different sequences were not performed in an effort to validate this general recommendation).

Another paper in 1999 dealt with the importance of particular DHPLC analytical conditions described in an earlier report [25]. This newer report focused on mutations in BRCA1 and BRCA2 [26], and illustrates further the importance of highly accurate temperature selection noted earlier [27]. For one, this investigation applied prediction of analysis temperatures – the algorithms for predictions can be found at the Stanford DHPLC website (http://insertion.stanford.edu/melt.html) or were provided in a commercial DHPLC software package called WAVEMaker® from Transgenomic, (Omaha, NE). No details were provided for either algorithm. For 180 different mutations tested, two mutations could only be detected at a single temperature, thus demonstrating the critical nature of not only selecting the correct temperature, but also in the requirement at all times for column temperature

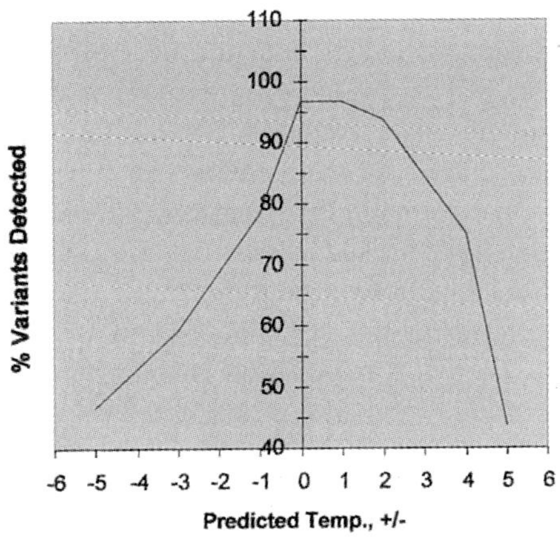

Figure 4.11. Sensitivity of DHPLC analyses to proper temperature selection.

accuracy. The authors also discussed the effect of moving PCR primers on the template when a particularly troublesome mutation was not yielding to detection, namely exon 5 of BRCA1. The authors asserted that these difficulties "...may have been due to the ability of the fragment to form a stable secondary cruciform structure", but do not provide supporting data in this regard. The initial exon 5 primers were those described in an earlier publication on SSCP analyses of BRCA1 [29]. These primers resulted in three out of four specific exon 5 mutations having indeterminate peak shapes. When the primers were shifted on the template to yield a different amplicon that possessed these mutations (no details were given as to how these new primers positions were selected), detection of one of these three mutations was considerably improved, and detection of another was somewhat improved. Detection of the third mutation was not appreciably improved.

There were two papers from one group that characterized mutation detection in BRCA1 by DHPLC, where comparisons were also made between DHPLC and cycle sequencing [30, 31], as well as SSCP [31]. Both of these studies demonstrated 100% sensitivity for the BRCA1 mutations confirmed by sequencing. Also, both studies deal with shifting PCR primer locations relative to prior recommendations [29] to gain DHPLC sensitivity, but do not describe the processes or procedures they used for repositioning these new primers. One of these papers draws some additional points worth noting [30]. For one, this paper provided results obtained with C-18 reverse phase columns having substrates of poly(styrenedivinylbenzene) (PS-DVB), as well as non-porous silica (with and without end-capping) (See Chapter 3). The PS-DVB substrate allowed for more than 5,000 successful analyses. The silica substrate without end-capping allowed only 10 successful analyses, while the end-capped substrate provided about 500 successful analyses. This paper, in keeping with another publication on BRCA1 mutation detection [28], points out the need for highly accurate temperature control by showing a SNP that could only be detected at one temperature. Lastly, this paper contemplates the potential for genotyping based on signal peak shape differences.

Two additional publications gave more detailed examinations of the effect of fragment melting domains on DHPLC mutation detection. One report dealt with placing the PCR primers on the template such that a single melting domain was generated after amplification [32]. This strategy led to the selection of a single temperature for the entire amplicon, where the analysis temperature was selected so that the entire fragment was essentially on the verge of complete melting. The investigators hypothesized that this approach led to 100 % sensitivity achieved for numerous mutations in the first 20 exons of TSC2. This study also compared the performance of DHPLC to SSCP (63 % sensitivity) and heteroduplex analysis (54 % sensitivity). A second report mentioned earlier in Section 4.2.2 on temperature prediction [7] deals with a single 279 bp fragment within the HPRT gene. This fragment possessed clearly differentiable "low-melting" and "high-melting" domains, with 20 distinct mutations dispersed across these separate domains. It was shown that when a mutation was present within the low-melting domain, the degree of resolution between heteroduplices and homoduplices was significantly greater than when the mutation was present within the high-melting domain. The authors also pointed out the value of reproducibly generating highly resolved mutant species, in that it allows for concentration and purification of rare mutants from a higher concentration of non-mutated DNA. Following on this theme, the ability to physically collect well resolved mutant heteroduplex species from homoduplices was demonstrated in a separate paper [33]. This paper also showed that the collected fragments could be subjected to further processes, where these collected fragments were subjected to additional PCR. This general increase in understanding DHPLC related phenomena resulted in the high levels of confidence required for performing highly ambitious studies that involve high-throughput scanning of large numbers of bases in sizable populations. Example of this type of study include BRCA2 [34], significant portions of chromosome 5 [35] and chromosome 21 [36], coding and additional non-coding regions of 106 different candidate disease genes [6], genome-wide loci in A. thaliana [37], and a functional region of the Y chromosome in over 600 men [38].

DHPLC publications continued apace in 2000 and the first part of 2001, where investigators continue to extend the boundaries of what can be performed by DHPLC, and in continual efforts to enhance its productivity. In one publication the authors were particularly interested in determining mutations within GC-rich sequences, which is a perennial challenge for any mutation detection scheme. These authors determined that their mutation detection sensitivity was 97 % within Notch3 exonic sequences, where GC content ranged from 57.8 % to 67.7 % [39]. In other publications, including the publication just cited, investigators were very interested in the ability of DHPLC to discriminate (or "score") particular mutations on the basis of distinct peak shapes, thus providing the simultaneous capability to genotype the sample in question and thus forego subsequent sequence analysis. While certain reports regarding the analysis of BRCA1 and BRCA2 demonstrate that this is possible in certain circumstances [40], other reports indicate that this is not the case, and that identical traces are obtained for different mutations such as in EXT1 and EXT2 mutation analyses [41], as well as the report cited earlier regarding Notch3 [39]. In the theme of "scoring" particular mutants by DHPLC, a report was also given for the detection of two putative mutations from two distinct codons (63 and 282) in the HFE gene, whereby two separate PCR products harboring these codons were injected for simultaneous analysis by DHPLC. The authors reported 100 % concordance with PCR/restriction enzyme detection [42]. Also, an approach to DHPLC detection of cytosine methylation via standard bisulphite treatment of starting genomic DNA was described [43].

In other reports, the capability of DHPLC to detect low concentrations of mutant sequences were central in a number of publications. Investigators used DHPLC for scanning pooled genomes in A. thaliana for detection of induced mutations, as part of the investigators' larger process referred to as TILLING (Targeting Induced Local Lesions IN Genomes). These investigators were capable of scanning as many as 20 pooled sequences at one time [44]. Another ambitious publication focused on DHPLC's capability to simultaneously scan for mutations while quantifying the level of the mutations. This was reported for the analysis of the entire mtDNA genome by DHPLC, where the authors reported a detection limit of 0.5 % for the A8344G mutation [45]. A recently issued PCR/DHPLC patent described how it is not necessary to actually observe the low concentration of mutant material as a peak with the primary detection scheme, since DNA Chromatography allows for physical collection of the putative mutant sequences along with its subsequent PCR amplification for final detection [46].

One of the most exciting and intriguing developments within DHPLC has been the recent development of capillary monolith columns with mass spectrometric detection, the benefits of which were described in the previous chapter on DNA separations in general. This work showed that DNA Chromatography is capable of achieving the gains observed previously for protein separations with capillary monolith columns (47), and that this approach holds the potential in the authors' opinion for the development of highly multiplexed capillary arrays [48]. This work was extended to DHPLC analysis of mutant sequences [49], where these authors subsequently used a four-color laser-induced fluorescence detector for the simultaneous analysis of four amplicons that each carry a different fluorophore [50]. The authors point out that the fluorescent labels' differential effect on relative adsorption to the column can influence the final analysis temperature, since each amplicon may be at a different acetonitrile concentration at the point of denaturation from the column.

4.4 Conclusions

In the five years since DHPLC was introduced to the research community, it has moved from being performed by a very small number of practitioners to being performed in hundred of laboratories, thus providing DHPLC with an increasingly mainstream status. This change in status is due in large part to the technique's commercialization, but primarily to the efforts of many researchers the world over who have put tremendous effort into making the technique as widely accepted as it is today. Enhanced understanding over the past five years of the fundamental chemical processes within DHPLC separations will lead to great strides in the foreseeable future in the areas of:

- Enhanced detection accuracy
- Enhanced sensitivity for minority concentrations of mutants
- Enhanced automation
- Improved methods for ensuring the accuracy of the oven temperature
- Greater simplification of analyses through miniaturization and improved prediction of analytical conditions
- Reduced cost-per-analysis
- Multiplexing strategies (on-column, multicolumn)
- New detection schemes (mass spectrometry)
- Data processing and signal interpretation

With the gains made in the past five years, and these anticipated gains in the next five years, we will undoubtedly see DHPLC become a standard tool for genetic variant detection within many, many more laboratories. This serves as a testament to the vision of the original inventors of this technology, as well as the technology's early adopters who saw the promise of a highly flexible, sensitive, and automated approach to mutation analysis.

Some of the common questions and answers for performing DHPLC analysis are listed in Appendix 3 at the end of the book.

References

1. P. A. Underhill, L. Jin, R. Zemans, P. J. Oefner, and L. L. Cavalli-Sforza, A pre-Columbian Y chromosome-specific transition and its implications for human evolutionary history, Proc. Natl. Acad. Sci. USA, 93, 196, 1996.
2. R. G. H. Cotton, Mutation Detection, Oxford University Press: Oxford, UK, 1997.
3. D. Poland, Recursion relation generation of probability profiles for specific-sequence macromolecules with long-range correlations, Biopolymers, 13, 1859, 1974.
4. M. Fixman and J. J. Friere, Theory of DNA melting curves, Biopolymers, 16, 2693, 1977.
5. K. J. Breslauer, R. Frank, H. Bloecker and L. A. Marky, Predicting DNA duplex stability from the base sequence, Proc. Natl. Acad. Sci. USA, 83, 3746, 1986.
6. M. Cargill, D. Altshuler, J. Ireland, P. Sklar, K. Ardlie, N. Patil, C. R. Lane, E. P. Lim, N. Kalyanaraman, J. Nemesh, L. Ziaugra, L. Friedland, A. Rolfe, J. Warrington, R. Lipshutz, G. Q. Daley and E. S. Lander, Characterization of single-nucleotide polymorphisms in coding regions of human genes, Nature Genetics, 22, 231, 1999.
7. T. R. Skopek, W. E. Glaab, J. J. Monroe, K. L. Kort and W. Schaefer, Analysis of sequence alterations in a defined DNA region: Comparison of termperature-modulated heteroduplex analysis and denaturing gradient gel electrophoresis, Mutation Res., 430, 13, 1999.
8. WIPO PCT Publication WO/99-07899.
9. D. T. Gjerde and P. D. Taylor, System and method for automated matched ion polynucleotide chromatography, PCT WO01/46687, 2001.
10. D. T. Gjerde, P. D. Taylor and R. Haefele, Denaturing multi ion polynucleotide chromatography for detecting mutations PCT WO 99/07899, 1999.
11. G. Narayanaswami and P. D. Taylor, Improved efficiency of mutation detection by high-performance liquid chromatography using modified primers and hybridization procedures, Gen. Test., 5, 9, 2001.
12. K. Khrapko, J. S. Hanekamp, W. G. Thilly, A. Belenkii, F. Foret, and B. Karger, Constant denaturant capillary electrophoresis (CDCE): a high resolution approach to mutational analysis, Nucleic Acids Research, 22, 364, 1994.
13. N. R. Soman and G. N. Wogan, Activation of the c-Ki-ras oncogen in aflatoxin B1-induced hepatocellular carcinoma and adenoma in the rat: detection of denaturing gradient gel electrophoresis, Proc. Natl. Acad. Sci. USA, 90, 2045, 1993.
14. K. H. Hecker, P. D. Taylor and D. T. Gjerde, Mutation detection by denaturing DNA chromatography using fluorescently labeled PCR products, Anal. Biochem., 272, 156, 1999.
15. K. R. Tindall and T. A. Kunkel, Fidelity of DNA synthesis by the Thermus aquaticus DNA polymerase, Biochemistry, 27, 6008, 1988.
16. K. A. Eckert and T. A. Kunkel, The fidelity of DNA polymerases used in the polymerase chain reaction, PCR: A Practical Approach, IRL Press/Oxford University Press, Oxford: UK, 1991.
17. W. Xiao and P. J. Oefner, Denaturing High-Performance Liquid Chromatography: A Review, Human. Mut., 17, 439, 2001.
18. R. A. Ophoff, G. M. Terwindt, M. N. Vergouwe, R. van Eijk, P. J. Oefner, S. M. G. Hoffman, J. E. Lamerdin, H. W. Mohrenweiser, D. E. Bulman, M. Ferrari, J. Haan, D. Lindhout, G.-J. B. van Ommen, M. H. Hofker, M. D. Ferrari and R. R, Frants, Familial hemiplegic migraine and episodic ataxia type-2 are caused by mutations in the Ca^{2+} channel gene CACNL1A4, Cell, 87, 543, 1996.
19. P. A. Underhill, L.. Jin, A. A. Lin, S. Q. Mehdi, T. Jenkins, D. Vollrath, R. W. Davis, L. L. Cavalli-Sforza and P. J. Oefner, Detection of numerous Y chromosome polymorphisms by denaturing high-performance liquid chromatography, Genome Res., 7, 996, 1997.
20. W. Liu, C. D. James, L. Frederick, B. E. Alderete and R. B. Jenkins, PTEN/MMAC1 mutations and EGFR amplifications in glioblastomas, Cancer Res., 57, 5254, 1997.
21. W. Liu, D. I. Smith, K. J. Rechtzigel, S. N. Thibodeau and C. D. James, Denaturing high-performance liquid chromatography (DHPLC) used in the detection of germline and somatic mutations, Nucl. Acids Res., 26, 1396, 1998.

22. A. Yokomizo, D. J. Tindall, H. Drabkin, R. Gemmill, W. Franklin, P. Yang, K. Sugio, D. I. Smith and W. Liu, PTEN/MMAC1 mutations identified in small cell, but not in non-small cell lung cancers, Oncogene, 17, 475, 1998.
23. M. Mai, A. Yokomizo, C. Qian, P. Yang, D. J. Tindall, D. I. Smith and W. Liu, Activation of p73 silent allele in lung cancer, Cancer Research, 58, 2347, 1998.
24. M. C. O'Donovan, P. J. Oefner, S. C. Roberts, J. Austin, B. Hoogendoorn, C. Guy, G. Speight, M. Upadhyaya, S. S. Sommer and P. McGuffin, Blind analysis of denaturing high-performance liquid chromatography as a tool for mutation detection, Genomics, 52, 44, 1998.
25. A. Kuklin, K. Munson, D. Gjerde, R. Haefele and P. Taylor, Detection of single-nucleotide polymorphisms with the Wave DNA Fragment Analysis System, Genetic Testing, 1, 201, 1998.
26. W. Liu, P. J. Oefner, C. Qian, R. S. Odom and U. Francke, Denaturing HPLC-identified novel FBN1 mutations, polymorphisms and sequence variants in Marfan Syndrome and related connective tissue disorders, Genetic Testing, 1, 237, 1998.
27. A. C. Jones, J. Austin, N. Hansen, B. Hoogendoorn, P. J. Oefner, J. P. Cheadle and M. C. O'Donovan, Optimal temperature selection for mutation detection by denaturing HPLC and comparison to single-stranded conformation polymorphism and heteroduplex analysis, Clin. Chem., 45, 1133, 1999.
28. T. Wagner, D. Stoppa-Lyonnet, E. Fleischmann, D. Muhr, S. Pages, T. Sandberg, V. Caux, R. Moeslinger, G. Langbauer, A. Borg, and P. Oefner, Denaturing high-performance liquid chromatography detects reliably BRCA1 and BRCA2 mutations, Genomics, 62, 369, 1999.
29. L. S. Friedman, E. A. Ostermeyer, C. I. Szabo, P. Dowd, E. D. Lynch, S. E. Rowell and M. C King, Confirmation of BRCA1 by analysis of germline mutations linked to breast and ovarian cancer in ten families, Nature Genetics, 8, 399, 1994.
30. N. Arnold, E. Gross, U. Schwarz-Boeger, J. Pfisterer, W. Jonat and M. Kiechle, A highly sensitive, fast, and economical technique for mutation analysis in hereditary breast and ovarian cancer, Human Mutation, 14, 333, 1999.
31. E. Gross, N. Arnold, J. Goette, U. Schwarz-Boeger and M. Kiechle, A comparison of BRCA1 mutation analysis by direct sequencing, SSCP and DHPLC, Human Genetics, 105, 72, 1999.
32. Y. S. Choy, S.L Dabora, F. Hall, V. Ramesh, Y. Niida, D. Franz, J. Kasprzyk-Obara, M. P. Reeve and D. J. Kwiatkowski,Superiority of denaturing high-performance liquid chromatography over single-stranded conformation and conformation-sensitive gel electrophoresis for mutation detection in TSC2, Annals Hum. Genet., 63, 383, 1999.
33. A. Kuklin, K. Munson, P. Taylor and D. Gjerde, Isolation and analysis of amplified cDNA fragments during detection of unknown polymorphisms with temperature modulated heteroduplex chromatography, Molec. Biotech., 11, 257, 1999.
34. T. M. U. Wagner, K. Hirtenlehner, P. Shen, R. Moeslinger, D. Muhr, E. Fleischmann, H. Concin, W. Doeller, A. Haid, A. H. Lang, P. Mayer, E. Petru, E. Ropp, G. Langbauer, E. Kubista, O. Scheiner, P. Underhill, J. Mountain, M. Stierer, C. Zielinski and P.Oefner, Global sequence diversity of BRCA2: analysis of 71 breast cancer families and 95 control individuals of worldwide populations, Human Molecular Genetics, 8, 413, 1999.
35. M. Giordano, P. J. Oefner, P. A. Underhill, L. L. Cavalli-Sforza, R. Tosi and P. M. Richiardi, Identification by denaturing high-performance liquid chromatography of numerous polymorphisms in a candidate region for multiple sclerosis susceptibility, Genomics, 56, 247, 1999.
36. L. Jin, P. A. Underhill, V. Doctor, R. W. Davis, P. Shen, L. C. Cavalli-Sforza and P. J. Oefner, Distribution of haplotypes from a chromosome 21 region distinguishes multiple prehistoric human migrations, Proc. Natl. Acad. Sci. USA, 96, 3796, 1999.
37. R. J. Cho, M. Mindrinos, D. R. Richards, R. J. Sapolsky, M. Anderson, E. Drenkard, J. Dewdney, T. L. Reuber, M. Stammers, N. Federspiel, A. Theologis, Y. H. Yang, E. Hubbell, M. Au, E. Y. Chung, D. Lashkari, B. Lemieux, C. Dean, R. J. Lipshutz, F. M. Ausubel, R. W. Davis and P. J. Oefner, Genome-wide mapping with biallelic markers in Arabidopsis thaliana, Nature Genetics, 23, 203, 1999.
38. C. Sun, H. Skaletsky, B. Birren, K. Devon, Z. Tang, S. Silber, R. Oates and D. C. Page,

An azoospermic man with a de novo point mutation in the Y-chromosomal gene USP9Y, Nature Genetics, 23, 429, 1999.
39. J. L. Escary, M. Cecillon, J. Maciazek, M. Lathrop, E. Tournier-Lasserve and A. Joutel, Evaluation of DHPLC analysis in mutational scanning of Notch3, a gene with a high G-C content, Human Mutation, 16, 518, 2000.
40. E. Gross, N. Arnold, K. Pfeifer, K. Bandick and M. Kiechle, Identification of specific BRCA1 and BRCA2 variants by DHPLC, Human Mutation, 16, 345, 2000.
41. C. Dobson-Stone, R. D. Cox, L. Lonie, L Southam, M. Fraser, C. Wise, F. Bernier, S. Hodgson, D. E. Porter, A. H. R. W. Simpson and A. P. Monaco, Comparison of fluorescent single-strand conformation polymorphism analysis and denaturing high-performance liquid chromatography for detection of EXT1 and EXT2 mutations in hereditary multiple exostoses, European J. Hum. Genet., 8, 24, 2000.
42. S. G. Kaler, J. M. Devaney, E. L. Pettit, R. Kirshman and M. A. Marino, Novel method for molecular detection of the two common hereditary hemochoromatosis mutations, Genetic Testing, 4, 125, 2000.
43. D. Deng, G. Deng, J. Zhou and H. Xin, Detection of CpG methylations in human mismatch repair gene HMLH1 promoter by denaturing high-performance liquid chromatography (DHPLC), Chin. J. Cancer Res., 12, 188, 2000.
44. C. M. McCallum, L. Comai, E. A. Greene and S. Henikoff, Targeted screening for induced mutations, Nature Biotech., 18, 455, 2000.
45. B. J.C . van den Bosch, R. F.M de Coo, H. R. Scholte, J. G. Nijland, R. van den Bogaard, M. de Visser, C. E. M. de Die-Smulders and H. J. M. Smeets, Mutation analysis of the entire mitochondrial genome using denaturing high-performance liquid chromatography, Nucl. Acids Res., 20, e89, 2000.
46. J. L. Sklar, D. T. Gjerde, K. A. Lamb and C. P. Hanna, Method for detection of DNA by denaturing matched ion polynucleotide chromatography (DMIPC) and PCR, U. S. Pat. 6,218,153, 2001.
47. I. Gusev, X. Huang and C Horvath, Capillary columns with in situ formed monolithic packing for micro high-performance liquid chromatography and capillary electrochromatography, J. Chromatogr., 855, 273, 1999.
48. A. Premstaller, H. Oberacher and C.G Huber, High performance liquid chromatography-electrospray ionization mass spectrometry of single- and double-stranded nucleic acids using monolithic capillary columns, Anal. Chem., 72, 4386, 2000.
49. C. G. Huber, A. Premstaller, W. Xiao, H. Oberacher, G. K Bonn and P. J. Oefner, Mutation detection by capillary denaturing high-performance liquid chromatography using monolithic columns. J. Biochem. Biophys. Methods, 47, 5-19, 2001.
50. W. Xiao, D. Stern, M. Jain, C. G. Huber and P. J. Oefner, Multiplex capillary denaturing high-performance liquid chromatography with laser-induced fluorescence detection, BioTechniques, 30, 1332, 2001.

5
Size Based Separations

5.1
Introduction

The earliest reports related to DNA Chromatography on non-porous reverse-phase polymeric media dealt largely with its remarkable ability to perform true size-based separations of double-stranded DNA in a rapid manner. This technology was also demonstrated to have the ability to perform single-stranded DNA separations in a nearly size-based manner (i. e. partial dependence upon base composition). While these early reports provided a great deal of information on the fundamental aspects of these separations, they also introduced beneficial aspects of DNA Chromatography to applications that require size-based separations. In more recent years, these beneficial aspects have been directed towards an increasingly larger selection of applications. These two broad areas of DNA Chromatography for size-based separations – fundamental developments and applications – are discussed below. Also discussed are some factors affecting calibration of the system for size determination.

5.2
Fundamental Developments

Before there were reports of true size-based separations with currently applied DNA Chromatography techniques, there were a number of reports on high-resolution separation of various types of nucleic acids. The first of several papers by G. Bonn, C. Huber, P. Oefner and other coworkers appeared in 1993 [1]. The paper clearly demonstrates the fundamental gains made by utilizing non-porous reverse-phase polymeric media. The non-porous polymeric media, along with the highly monodisperse nature of the particles, results in minimization of the diffusion paths – thus providing extremely high resolution with very short analysis times (not to mention creating a robust media that is resistant to a broad range of pH conditions). Furthermore, it was demonstrated that when the particles were alkylated with octadecyl groups, the aromatic rings of the polymeric media are shielded from the hydrophobic regions within the nucleotides. This provided additional resolution for the separation of fragments larger than 200 bp. The investigators also examined the effects of pairing ion concentration (triethylammonium acetate), column temperature, acetonitrile concentration, eluent flow rate, and mass of material injected. The authors adopted the optimized separation conditions for the analysis of RT-PCR products derived from the multi drug resistance gene.

A subsequent paper by the same group focused on the speed with which various biopolymers (nucleic acids and proteins) could be analyzed with these conditions [2]. It was noted that this was the first published report of ion-pair reverse-phase HPLC separation of oligonucleotides and DNA fragments in less than 60 seconds, and also showed a very realistic and practical separation of a 20-mer oligonucleotide from a 19-mer "single-base failure" oligonucleotide in less than 30 seconds. Furthermore, 10 replicate analyses of the same unpurified PCR product were completely analyzed in 20 minutes.

Once these authors established the general utility and validity of this approach to nucleic acid analyses, their attention was focused onto very specific categories of nucleic acid analyses. The first study of this nature involved the analysis of oligonucleotides [3]. While exceptionally high resolution and rapid analysis times were demonstrated for oligonucleotides, the authors also demonstrated that equal length oligonucleotides experience different retention times as a result of differences in base content. For example, it was shown that for a 22-mer oligonucleotide where the 3′ nucleotide is varied, the elution order was C<G<A<T. This elution order was consistent with the degree of hydrophobicity provided by each base. This general dependence on base composition leads to near-size-based separations for single stranded DNA in general, which is observed in subsequent applications of DNA Chromatography.

Nucleic acids in general often carry fluorescent tagging moieties, so these authors also examined how they impact nucleic acid separations by DNA Chromatography [4]. In general the authors demonstrated that the presence of fluorescent tags did not inherently degrade the performance of DNA Chromatography separations. However, it is notable that the fluorescent tags do have an influence on the separation as they contribute to an overall increase in hydrophobicity, and that different tags have different degrees of influence on hydrophobicity. The authors examined the effect of the pairing ion's relative hydrophobicity in an attempt to elucidate fundamental retention mechanisms, which allows for explanation of the variable retention behaviors for different fluorescent dyes. Demonstration of the usefulness of fluorescent tags in DNA Chromatography was achieved through analysis of multiplexed RT-PCR fragment pools (results compared favorably to those obtained with polyacrylamide gel electrophoresis), as well as the analysis of fluorescently labeled hybridization reactions.

The authors of these earlier reports subsequently published a paper that specifically addresses the topic of true size-based separations of double stranded DNA by DNA Chromatography [5]. Factors such as reproducibility of substrate alkylation and column equilibration were examined for different plasmids digested with different restriction enzymes. These experiments unequivocally demonstrated highly reproducible size-based separations of double stranded DNA with high resolution and separation speed. This independence from sequence content is a radical departure from previous observations with anion-exchange chromatography, where the relative AT content exerted a significant impact on fragment retention.

The difference between anion-exchange and ion-pair reverse-phase separation mechanisms was the topic of the next two reports from these authors, specifically through the use of micropellicular stationary phases for near-size-based separations of oligonucleotides [6] and size-based separations of double-stranded DNA [7]. As the authors state, these stationary phases lack pores that are accessible to macromolecules, thus providing extremely rapid mass transfer between the stationary phase and the mobile phase (among other characteristics). In the case of oligonucleotide separations [6], the authors point out the clear superiority of micropellicular materials over silica-based materials, as well as demonstrating the enhanced column efficiency when the poly(styrene-divinylbenzene) material was alkylated vs the anion-exchange separation. This latter approach has the additional advantage of possessing fully volatile eluent components. Similar analyses were performed on double-stranded DNA [7], where the authors examined and compared column efficiency, retention characteristics, and independence from sequence content on size-based separations. While it was possible to make anion-exchange separations size dependent (based on the pool of fragments analyzed), the best performance/agreement was obtained with poly(styrene-divinylbenzene) alkylated with octadecyl groups. The authors also provide an excellent review of work performed by others regarding the use of micropellicular stationary phases for preparative fractionation, restriction analysis, analysis of PCR products, analysis of fluorescently labeled PCR products from HIV, detection of partial denaturation of PCR products (DHPLC, discussed in Chapter 4), as well as purification of DNA plasmids (nicked vs. supercoiled, linear vs. supercoiled).

5.3 Calibration

Size determination of unknown fragments depends on a reliable method of calibrating a column to determine the relationship between the retention time and the base pair length of eluted fragments. When electrophoresis is used, size calibration of a slab gel for size separation is accomplished by running a DNA ladder on each end lane of the gel bracketing the samples run in the lanes in between. This is to compensate for any differences in the gel being used from other gels and for variations within the slab itself.

Obviously this method cannot be used for DNA Chromatography since only one column is usually available. The most common method of column calibration is to use an external size standard (see Chapters 2 and 6 for discussions on internal and external standards). Figure 5.1 shows several separations of a DNA ladder and an enzymatic digest. In order for the use of an external standard to be successful, the separation must be reproducible from run to run as demonstrated in this figure. Retention times are reproducible from run to run or even from day to day. Therefore, DNA ladders or other standards can be used to calibrate column for the reten-

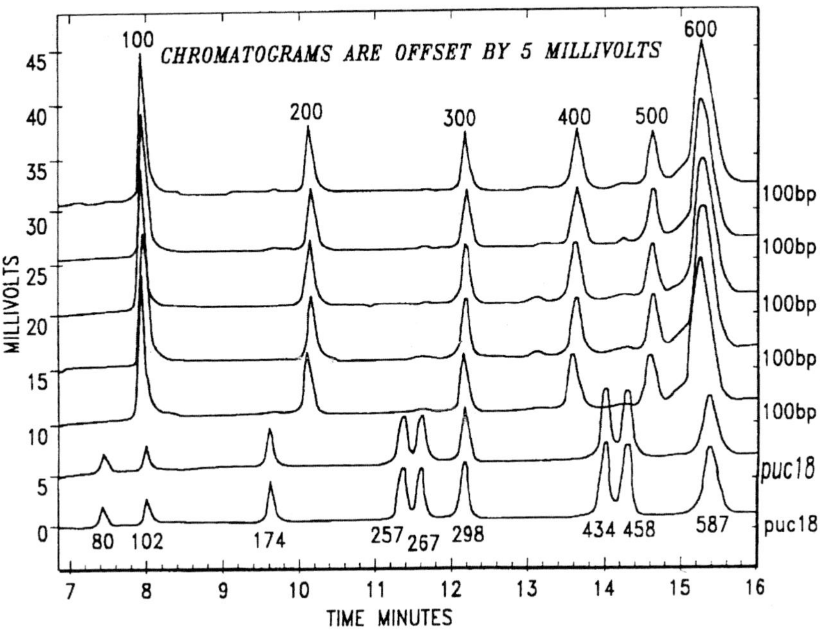

Figure 5.1. Separaton of DNA 1000 bp ladder and pUC 18 Hae III enzymatic digest (from Ref [8] with permission).

tion times relative to base pair length. However, the most accurate results are measured when the calibration run is performed either before or after the sample run.

A plot of the retention times of different fragment sizes vs. the amount of acetonitrile in the eluent being introduced into the chromatograph is shown in Figure 5.2. The plot is curved indicating that as the concentration of acetonitrile increases the eluting strength of the eluent also increases. This means when a linear gradient of acetonitrile is used, a curved calibration plot of fragment retention time vs. fragment size will be produced.

A common misconception is that this calibration plot should be linear. In fact a number of publications have plotted retention time vs. fragment size as a linear function and then tried to force a fit to a linear equation and calculate a linear correlation coefficient. It is possible to choose a nonlinear gradient that will give a linear retention time vs. size plot. Some software functions actually are actually designed to do this through knowledge of the separation characteristics of the system. But it is dangerous to assume one is using this type of eluent gradient. A plot should always be drawn using curve fitting and the calibration curve established.

Another potential problem in size calibration discovered by Gjerde and Taylor is the use of nicked DNA ladder as a size standard [8]. It turns out that many DNA ladders available on the market contain nicks (although improvements have been made since the work was first done). The term nick is a double stranded DNA fragment where in one of the stand is contiguous and the complimentary strand contain sat least one break where in the two adjacent bases are not covalently linked.

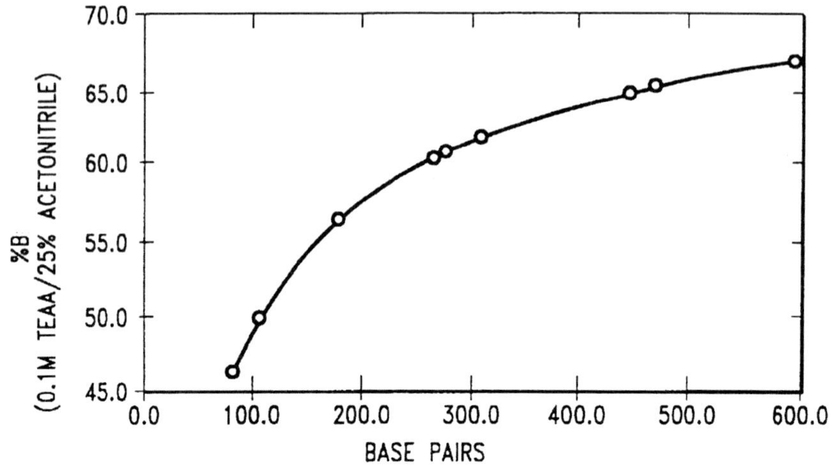

Figure 5.2. This plot of % B eluent containing 25 % acetonitrile vs. fragment retention time is curved indicating the strength of the eluent increases with concentration. This means that under most cases, calibration plots of fragment retention time vs. fragment size will also be curved. In some cases, a curved gradient can be formed to compensate for this behavior to give linear calibration (from Ref [8] with permission).

DNA ladders are mixture where the fragments comprise a defined range of base pair lengths and the fragments in the ladder different by a constant base pair increment. For example, a 1000 base pair ladder may contain DNA fragments which differ by 100 base increments over a range 100 to 1000 by i.e. 100, 200, 300, ... 1000. The accuracy of such ladders is of critical importance, since defective or impure ladders may lead to incorrect calculations.

Nicks in DNA fragments may occur for a variety of reasons. In the case of DNA ladders, it is likely that the longer DNA fragments are constructed by enzymatic ligation of the shorter strands. Some fragments may not be completely ligated (attached on strand but not the other). Or the ladder may contain a number of nicks on both sides.

The presence of nicks is shown in Figure 5.3. This is a separation of the same type of 1000 bp ladder as shown Figure 5.1 except this ladder does not show the expected pattern. The peak at just under 10 minutes retention time and to the left of the 200 bp fragment is a nicked 200 bp fragment. This conclusion was confirmed by collecting the nicked fragment and reinjecting the material under denaturing conditions at 80 °C. Under these conditions, an unnicked fragment would produce two single strands and two peaks (because the separation of single-stranded DNA is sequence dependent). A nicked fragment would be expected to produce more than two fragments. If the nick is only one side or the other then 3 fragments would be produced under denaturing conditions. In this case denaturation produces 5 fragments as shown in Figure 5.4 indicating the collected fragment is a mixture either one side or the other nicked. Six fragments species are expected but 2 of the peaks are coeluting giving only 5 peaks

[8]. The mass of peaks after the 200 bp fragment in Figure 5.3 indicates that virtually all of the fragments in the ladder contain nicks making the standard completely usable for calibration.

There is another interesting aspect. Nicks can also be formed when an enzyme which recognizes a base pair mismatch in a heteroduplex, binds within the vicinity of a mutation and cleaves one strand of the DNA duplex while leaving the other strand intact. Many such enzymes are known are the basis of one form of mutation detection. In DNA Chromatography, the nicked mutation could be detected under either non denaturing or denaturing conditions or a combination of the both for confirmation purposes. References 9 and 10 describe many of these enzymes.

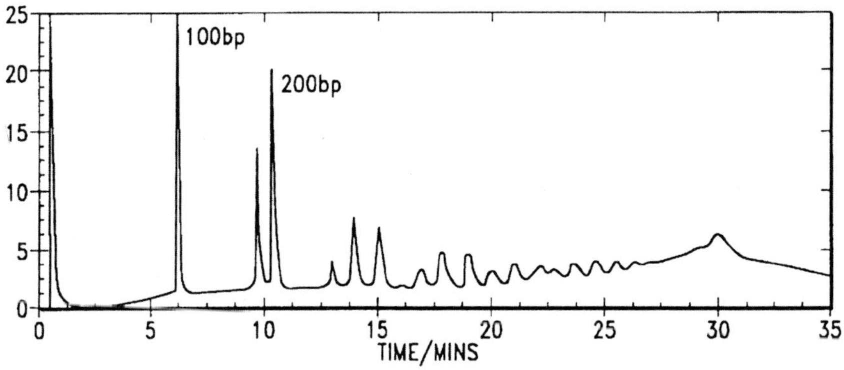

Figure 5.3. Separation of defective DNA ladder containing nicks (from Ref. [8] with permission).

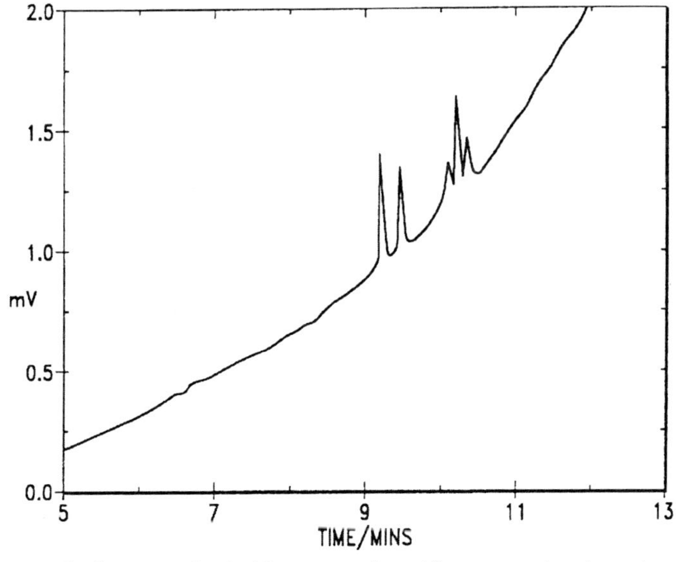

Figure 5.4. Reinjection of a fragment of nicked fragment collected from separation shown in Figure 5.3 (from Ref. [8] with permission).

5.4 Applications

Some of the earliest reports on size-based separations in DNA Chromatography also demonstrated its practical application. Since that time, additional reports have been published that clearly demonstrate the utility of size-based and near-size-based DNA Chromatography separations to a variety of analyses.

5.4.1 Primer extension and near-size-based separations

Near-size-based separations by DNA Chromatography have been applied for the purposes of genotyping through the analysis of primer extension products. The first report demonstrated primer-extension genotyping through relative sizing of single-stranded DNA [11]. Sites that harbored the SNP requiring genotyping were amplified by PCR, and primers were annealed adjacent to the polymorphic site (5′ to the SNP). The primer extension products were then generated in the standard manner. To analyze these extension products, the entire reaction mixture was loaded onto the column at 75 °C. This high column temperature completely denatured the extended primer from the host template, which allowed for direct near-sizing of the extension products on the column (and hence determination of the genotype). Accurate genotyping was demonstrated for a SNP in the proneurotensin gene, as well as two SNPs in the 5HT2a gene. The authors pointed out the ability to analyze more than one group of primer extension products at a time (thus improving throughput), as well as using as little as 10 ng PCR product with 0.1 units of Thermo-Sequenase for primer extension (thus decreasing cost per analysis).

This general approach has been applied in subsequent reports for the analysis of pooled genomes for determining allele frequencies within populations [12–15], with one report of incorporating fluorescent dye-labeled nucleotide terminators for identification of the polymorphic base(s) [15]. By first pooling the genomic samples from samples prior to PCR and primer extension, the procedure is simplified since it is the *frequency* of the allele within the population that is of interest, such as in family-based association studies. Authors have found that allele frequency differences generally on the order of only a few percent or less can be determined by these means.

Primer extension product analysis by DNA Chromatography has also been applied in circumstances where pooling is not an option, such as with molecular diagnosis of disease. Two separate groups have reported on the determination of two clinically significant polymorphisms in the HFE gene for diagnosis of hemochromatosis [16, 17]. Both groups of authors demonstrate that all 4 nucleotide terminators lead to different elution times depending upon which terminator is incorporated. Both groups also demonstrate the ability to analyze primer extension products from more than one locus at a time, thus enhancing throughput. Examples were given for determining the polymorphisms present at two loci at the same time (codons 63 and 282)

An approach that does not entail primer extension but bears some similarities to prior analyses is the analysis of completely denatured PCR products of sufficiently short length [18]. Just as different nucleotide terminators incorporated via primer extension leads to different elution times, different native nucleotides within a fully denatured PCR product ($< \sim 100$ nucleotides in length) will lead to different elution times, thus providing the basis for rapid, single-step genotyping procedure. This was partially demonstrated by accurately typing all possible transitions and transversions with the exception of C→G.

5.4.2
LOH and other size-based genotyping techniques

Size-based separations by DNA Chromatography represent a straightforward means of determining the loss-of-heterozygosity (LOH) that occurs as a result of chromosomal instability within molecular carcinogenesis. By judiciously selecting primer sequences that lead to PCR amplification of predetermined markers, it is possible to determine the ratio of alleles in a size-based manner. This has been demonstrated through amplification of dinucleotide microsatellite markers as well as amplification of wild-type and retroviral insertions of the TSC-2 gene in Eker rat colonies [19, 20]. Integrated peak areas for each PCR product determines the sought allele ratio for determination of LOH.

There have been a number of other reports related to genotyping procedures that require true size-based separations by DNA Chromatography. Investigators successfully genotyped BALB/c mouse colonies based on a 4 bp deletion, and were also able to resove the resultant heteroduplices generated in the process [21]. Investigators also examined various analytical parameters for accurate, DNA Chromatography-based typing of numerous short tandem repeats (STRs) of various lengths [22, 23], along with quantitative collection of alleles for subsequent confirmatory sequencing [22].

5.4.3
Size Based Purification Procedures

True size-based separations have been used to achieve highly selective DNA purification procedures. Specific PCR products have been size-separated and subsequently collected as a precursor to cloning procedures [24, 25] It was shown that these separation procedures not only automate otherwise labor-intensive gel electrophoresis procedures, but also removed primer-dimers and other non-specific amplification products that would otherwise have been cloned. Similar procedures were applied to more complex pools of PCR products generated by selective AFLP procedures for gene expression profiling [26]. Fractions were collected every 30 seconds over a 50 minute separation, and were subsequently sequenced to determine each fragment's identity.

References

1. C. G. Huber, P. J. Oefner, E. Preuss and G. K. Bonn, High-resolution liquid chromatography of DNA fragments on non-porous poly(styrene-divinylbenzene) particles, Nucleic Acids Res., 21, 1061, 1993.
2. C. G. Huber, P. J. Oefner and G. K. Bonn, Rapid analysis of biopolymers on modified non-porous polystyrene-divinylbenzene particles, Chromatographia, 37, 653, 1993.
3. C. G. Huber, P. J. Oefner and G. K. Bonn, High-resolution liquid chromatography of oligonucleotides on nonporous alkylated styrene-divinylbenzene, Anal. Biochem., 212, 351, 1993.
4. P. J. Oefner, C. G. Huber, F. Umlauft, G.-N. Berti, E. Stimpfl and G. K. Bonn, High-resolution liquid chromatography of fluorescent dye-labeled nucleic acids, Anal. Biochem., 223, 39, 1994.
5. C. G. Huber, P. J. Oefner and G. K. Bonn, Rapid and accurate sizing of DNA fragments by ion-pair chromatography on alkylated nonporous poly(styrene-divinylbenzene) particles, Anal. Chem., 67, 578, 1995.
6. C. G. Huber, E. Stimpf, P. J. Oefner and G. K. Bonn, A comparison of micropellicular anion-bxchange and reversed-phase stationary phases for HPLC analysis of oligonucleotides, LC-GC Mag., 14, 114, 1996.
7. C. G. Huber, Micropellicular stationary phases for high-performance liquid chromatography of double-stranded DNA, J. Chromatography A, 806, 1, 1998.
8. D. T. Gjerde and P. D. Taylor, Analysis of nicked DNA by matched ion polynucleotide chromatography, U. S. Pat. 6,187,539: 2001.
9. J. G. Chirikjian and B. G. Collier, Oscillating signal amplifier for nucleic acid detection, U. S. Pat 5,763,178: 1998.
10. Molecular Cloning, 2nd ed. Sambrook et al. Eds. Cold Spring Harbor Laboratory Press: 1989.
11. B. Hoogendoorn, M. J. Owen, P. J. Oefner, N. Williams, J. Austin and M. C. O'Donovan, Genotyping single nucleotide polymorphisms by primer extension and high performance liquid chromatography, Hum. Genet., 104, 89, 1999.
12. B. Hoogendoorn, N. Norton, G. Kirov, N. Williams, M. L. Hamshere, G. Spurlock, J. Austin, M. K. Stephens, P. R. Buckland, M. J. Owen and M. C. O'Donovan, Cheap, accurate and rapid allele frequency estimation of single nucleotide polymorphisms by primer extension and DHPLC in DNA Pools, Hum. Genet., 107, 488, 2000.
13. J. K. Wolford, D. Blunt, C. Ballecer and M. Prochazka, High-throughput SNP detection by using DNA pooling and denaturing high performance liquid chromatography (DHPLC), Hum. Genet., 107, 483, 2000.
14. M. Giordano, M. Mellai, B. Hoogendoorn and P. Momigliano-Richiardi, Determination of SNP allele frequencies in pooled DNAs by primer extension genotyping and denaturing high-performance liquid chromatography, J. Biochem. Biophys. Methods, 47, 101, 2001.
15. K. Kosaki, H. Yoshihashi, Y. Ohashi, R. Kosaki, T. Suzuki and N. Matsuo, Fluorescence-based DHPLC for allelic quantification by single-nucleotide primer extension, J. Biochem. Biophys. Methods, 47, 111, 2001.
16. J. M. Devaney, E. L. Pettit, S. G. Kaler, P. M. Vallone, J. M. Butler and M. A. Marino, Genotyping of two mutations in the HFE gene using single-base extension and high-performance liquid chromatography, Anal. Chem., 73, 620, 2001.
17. Q. Liang, P. A. Davis, B. H. Thompson and J. T. Simpson, High-performance liquid chromatography multiplex detection of two single nucleotide mutations associated with hereditary hemochromatosis, J. Chromatography A, 754, 265, 2001.
18. P. J. Oefner, Allelic discrimination by denaturing high-performance liquid chromatography, J. Chromatography B, 739, 345, 2000.
19. E. Kleymenova, S. Muga, S. Fischer and C. L. Walker, Application of high-performance liquid chromatography-based analysis of DNA fragments to molecular carcinogenesis, molecular carcinogenesis, 29, 51, 2000.
20. E. Kleymenova and C. L. Walker, Determination of loss of heterozygosity in frozen and paraffin embedded tumors by denaturing high-performance liquid chromatography, J. Biochem. Biophys. Methods, 47, 83, 2001.
21. A. Kuklin, A. P. Davis, K. H. Hecker, D. T. Gjerde and P. D. Taylor, A novel technique for rapid automated genotyping of DNA poly-

morphisms in the mouse, Mol. Cell Probes, 13, 239, 1999.

22. M. A. Marino, J. M. Devaney, J. K. Smith and J. E. Girard, Sequencing using capillary electrophoresis of short tandem repeat alleles separated and purified by high-performance liquid chromatography, Electrophoresis, 19, 108, 1998.
23. J. M. Devaney, J. E. Girard and M. A. Marino, DNA Microsatellite Analysis using ion-pair reversed-phase high-performance liquid chromatography, Anal. Chem., 72, 858, 2000.
24. K. H. Hecker, B. Turpie and A. Kuklin, Optimization of cloning efficacy by pre-cloning DNA fragment analysis, BioTechniques, 26, 216, 1999.
25. C. M. Shaw-Bruha and K. A. Lamb, Ion-pair reversed-phase HPLC (IP-RP-HPLC) approach facilitates sub-cloning of PCR Products and screening of recombinant colonies, BioTechniques, 28, 794, 2000.
26. L. Y. Wong, V. Belonogoff, V. L. Boyd, N. M. Hunkapiller, P. M. Casey, S.-N. Liew, K. D. Lazaruk and S. Baumhueter, General method for HPLC purification and sequencing of selected dsDNA gene fragments from complex PCRs generated during gene expression profiling, *BioTechniques*, 28, 776, 2000.

6
Purification of Nucleic Acids

6.1
Introduction

The classical method for high-resolution purification of nucleic acids is to perform a gel separation and then to use a razor blade or a scalpel to cut the fragment from the gel and then extract the fragment. Besides being time consuming this method can be inaccurate and have poor reproducibility and recovery. High-resolution, automated collection of nucleic acid fragments is a primary need within molecular biology research. This chapter describes the use of DNA Chromatography to collect nucleic acid fragments. The hardware, software and the procedures that are used depend on how the collected fragment will be used.

One of the primary considerations is to select conditions that will actually perform the separation. These conditions can be chosen to separate a range of fragment sizes or can be chosen to target a particular fragment. Once the separation is accomplished (the peak of interest is resolved), the process of collecting the material should not "undo" the separation process either through remixing of the bands or simply not collecting at the correct time. There are other considerations including keeping the bands free of contamination, recovery of material from the vial and use of the material. The matrix buffer of the collected material must be suitable for downstream processing. The general steps used for fragment collection are:

1. Configuration and plumbing of the instrument.
2. Cleaning and testing operation of the DNA Chromatograph.
3. Cleaning and testing operation of the fragment collector.
4. Calibration and performing the separation.
5. Collection of the material into vials.
6. Concentration and stabilization of the material (if needed).
7. Verification of collection, quantity and purity of material (if needed).
8. Use of the material for an application.

The word calibrate may have several meanings. In some cases it may mean to ensure that the pump, injector and detector are working properly. It may mean that the travel time between the detector cell and the fragment collector deposition probe has been correctly measured or that the conditions needed for collection of a particular fragment size are verified. Contamination and cleaning procedures are major issues for fragment purification. Possibly the most common problem is the tendency to flush to waste part or even the entire band that was to be collected and instead collect part of the peak, the adjacent fragment or nothing at all.

The configuration and properties of the instrument and fragment collector were discussed in Chapter 2. The following discussions describe the hardware, software, chemistry and procedures used for each of these steps for the collection of nucleic acid fragments. Examples of using DNA Chromatography to purify nucleic acids include applications of PCR [1–3], sequencing [4], cloning [5], oligonucleotides [2, 6], plasmids, STRs (short tandem repeats) [2, 7, 8], heteroduplex [9], and RNA [10]. Details of these applications are discussed in this chapter and other chapters in this book.

6.2 System Dead Volume

One of the most important features of an instrument is how much and what type dead volume the system contains. Dead volume is the volume within in the fittings, tubing, column, detector, injector, etc. where the eluent flows. While a channel is needed for the mobile phase to flow through the system and dead volume must exist, some types of dead volume can be detrimental.

There are two reasons for controlling dead volume for a system. The first reason is the reduction of peak broadening through the reduction of extra column effects. Of course, this holds true whether or not a fragment is to be collected. The other reason is improper fittings, extra fittings and various spurious channels can cause DNA to be retained on one injection and then be released on the next injection or several injections contaminating the peak being collected. The retention and release levels of DNA may not be apparent on the chromatogram because released DNA concentrations may be below the detection limits of the detector. Using tubing and fittings properly, keeping the number of frits and fittings in the system to a minimum and simply keeping the dead volume to a minimum will decrease or eliminate this effect. See discussion in Chapter 3 and below on cleaning on the role of metal contamination on the retention and release of nucleic acids.

The broadening effect of dead volume depends on its location in the system. Dead volume from the injector to the top of the column is least harmful. The injected nucleic acid will travel to the top of the column, be adsorbed and then wait for the correct concentration of acetonitrile in the mobile phase to remove the various fragment sizes from the column. The separation process has not yet started when the fragments are absorbed at the top of the column. They are in a tight band on the column bed regardless of the amount of dead volume before the column. For example, the bands are equally narrow if a small bore tubing or large bore tubing is used to connect the injector to the column.

Extra dead volume due to a poorly packed column is harmful to the separation but is beyond the control of the user. However, making proper use of the fittings (see Chapter 2) can minimize dead volume in the fittings connecting the column. Keeping the tubing inside diameter small and the length short should minimize dead volume between the column and detector cell. The effect of too large of a dead volume is shown as broad or tailing peaks in the chromatogram.

Dead volume effects cannot be measured after the detector cell. What may look on the display screen to be a well-resolved peak can actually be poorly resolved by the time the fragment is collected if the connection to the fragment collector is poor and the peak has broadened. This connection tubing should have a small inside diameter and a short length. The collector should be positioned right next to the detector cell. If the detector is not close to the bench top, the fragment collector should be positioned on a stand directly outside the detector cell to minimize the length.

6.3 Cleaning

Whether or not a system needs to be cleaned prior to collection of nucleic acid depends on the reason the nucleic acid is being collected. Some applications such as PCR are quite stringent. Contaminants having the correct primer sequence will be amplified along with the collected material. The amplification of the contaminant could be disproportional to what really exists in the sample. On the other hand, methods of collection for subsequent cloning do not have as stringent requirements (unless the material undergoes additional PCR prior to cloning).

If cleaning is necessary, then it is important to consider the parts of the system that require cleaning. Virtually any component that comes in contact with the DNA can be a source of contamination. The outside of the injection needle or deposition of probe or other tubing may need cleaning. There are routines for cleaning these, but frequently software can be programmed to perform multiple cleanings between runs. Cleaning can be an automated process with the attachment of valves and pumps [11]. Again, this can be part of the analysis process where the system is cleaned with every run or every several runs.

The column is one of the major sources for contamination. It has a high surface area due to the packing and frits where contaminants can adsorb. In addition the column has many channels, nooks and crannies where material can get physically trapped. Since contamination is most likely to occur at the top of the column, some cleaning procedures have the option of reversing the flow through the column. This can be accomplished by manually reversing the column or using a switching valve. Before any of this is done, the column manufacturer should be consulted.

The cleaning solution must not attack or deteriorate any of the surfaces with which it makes contact. Table 6.1 shows of list of cleaning solution reagents and the materials removed. Methanol is a polar solvent and can even remove polar complexes and salts, more nonpolar solvents such as acetonitrile are more effective in removing organics and other nonpolar materials [11–13].

Table 6.1. Cleaning solution reagents and materials removed (from Ref. [11] with permission)

Cleaning Solution Reagents	***Materials Removed***
Methanol, Acetonitrile, and organic solvents in general.	Organic contaminants, proteins, peptides, DNA residues, and genomic DNA.
Tetrasodium EDTA (0.1 – 500 mM) and other alkali metal and ammonium chelating agents. Heated solutions can be used.	Multivalent metal cations, DNA residues and genomic DNA.
Nitric acid (10 mM – 35 %). See passivation procedure Appendix 2. The column will have to be removed if a high concentration of nitric acid is used.	Metal ions, basic contaminants and the like. This treatment will also passivate the surface.
Sodium hydroxide, other alkali hydroxides and organic and inorganic bases	DNA residues, genomic DNA, proteins, peptides, and acidic contaminants
Enzyme, for example RNase or DNase.	Large DNA residues and genomic DNA. The enzyme may have to be cleaned from the column with another cleaning step.

Contaminants of metal ion DNA complexes may be neutral (nonpolar) and be removed with organic solvents. However, if there are metal ions in the system they should be removed with passivation procedures. Metal ions may appear not only from corrosion, but also from the sample itself. Many metals will hydrate (form hydroxides and then precipitate) at neutral pH and then become trapped within the system. Also, as noted in Chapter 3, many materials thought to be biocompatible may not actually be DNA compatible because of residual metals. Changing the tubing or another component in the system may introduce metal contaminants and cause DNA to be retained. If there is clear evidence that passivation cleaning is needed, then the column should be removed. Release of metals from the system during the passivation process could cause metal to become trapped on the column and harm it. Evidence that passivation is needed includes the appearance of an unexplained mystery peak, splitting peaks and disappearance of peaks. However, even small amounts of metal contamination could cause retention and release of nucleic acids [14–19].

High pH cleaning can be particularly useful because of the ability of the solution to degrade any nucleic acid that may be present. Tetrasodium EDTA serves a dual purpose because it can complex metal ions and also has a high pH to clean materials from the system. Solutions may be pumped through the system or simply injected depending on the amount of contamination. Solutions having a pH up to 13 and which can be heated by the system are particularly useful for cleaning by injection of cleaning solutions, although lower pHs are advised if the separation media is vulnerable to alkaline solutions, e. g. silica based particles and the like.

Larger nucleic acid contaminants such as genomic DNA cannot be removed from the column easily with normal solvent solutions. These and other nucleic acid contaminants can be removed by first digesting them into smaller fragments using an enzymes such as RNase or DNAZAP (Ambion, Austin TX). The digestion products are then removed by flowing solutions passed through the column.

6.4 Testing the Instrument Operation

Normally a size standard (usually pUC 18, Hae III) and mutation standard are tested on the instrument to confirm normal operation of the detector, injector, pumps, etc. and to ensure that the eluents were made correctly. The size and pattern of the peaks are compared with the chromatograms that are included with the new column. If the correct patterns are not seen, then the instrument eluent flow is first checked. After that, the eluent is checked and so on. Trouble-shooting guides are available from the instrument manufacturer.

Calibration of the fragment collector involves confirmation that the deposition probe positions itself in the center of each collection vial, does not splatter when a fluid is deposited into the vial, and does not deposit drops between vials. Also important is confirmation of the time lag between the detection of the peak and the actual collection. The exact procedure used depends on the manufacturer and is usually done at the manufacturing site. However, if tubing is changed, then this calibration must be repeated. One way to check to see if the time is operating correctly is to perform a collection with three vials. A method is programmed to deposit a complete peak into the middle vial and the other two vials containing fluid from each side of the peak. A known portion each vial is reinjected into the instrument to determine what was actually deposited into the vial. Of course, there should be no material in vials 1 and 3 and the amount of material in vial 2 should be at least 90 % recovery.

The recovery is calculated by using the ratio of the volume that was collected divided by volume that was reinjected. 100 % recovery is not usually achievable because not all of the sample taken up by the injection needle is actually injected into the eluent stream. A small portion of the injection volume flows to waste.

6.5 Calibration and Separation Conditions

6.5.1 Internal and External Calibration

Calibration for fragment collection must be performed to close tolerances for accurate collection of fragment sizes. Either internal standards or external size standards (usually an enzymatic digest) can be used. Internal standards are added to the sample and separated along with the fragments of interest. Both the sample and standard are subject to any changes in conditions that would result in any variation in the retention times of the fragments. A calibration plot (discussed later) is used to determine the time at which the fragment size of interest is eluting.

External size standards are injected separately from the sample. A mixture of fragment sizes bracketing the fragment size of interest is injected some time before the sample and then after the sample has been run and collected. The first standard run is used to perform the calibration to determine when to collect. The run after the sample is to confirm that the retention times for the target size did not change while the sample was being collected. External standards work quite well because the separation times are usually quite reproducible with deviation less than 1 %.

6.5.2
Isocratic Elution

Gradient-based separation methods are the primary means for performing separations. Generally, injections are made at low acetonitrile concentration to load the column (the fragments are adsorbed to the top of the column bed), then a gradient is employed to elute the fragments of interest. Finally, a sharp cleanoff wash is applied to remove any residual material (nucleic acids, proteins, surfactants, etc). The gradient is a powerful tool because one method can cover any size range of fragments without prior knowledge of the sample.

The starting concentration and slope of the acetonitrile gradient are selected to elute either a range of fragment sizes or a specific size in the desired time window. The steepness of the gradient chosen depends on how aggressively the separation is to be performed. A steep gradient is applied to achieve a fast separation. A more cautious gradient is applied if higher resolution is needed. Long shallow gradients are used for more difficult samples. Examples of this include samples of unknown fragment sizes, samples containing many different fragment types, and samples where fragments of interest are similar in size. As more information is acquired about a sample, the user generally adjusts the gradient to increase the separation speed. Increasing the resolution of a separation is more difficult.

In the work presented in this section, isocratic elution is introduced to increase peak resolution. It is based on using a constant concentration of eluent (isocratic elution). Isocratic elution is quite common in conventional HPLC because the process and equipment are simpler but isocratic elution is almost never used in DNA Chromatography. One reason is that the separations take longer because it is necessary to elute all the materials from the column before starting the next injection. But the primary reason is because it has been difficult to determine the conditions needed to accomplish the separation of interest. The sample fragments either elute off the column quickly because the acetonitrile concentration is too high, or the peaks never elute because the concentration is too low. This makes it difficult to use isocratic elution, because the peaks of interest either elute off quickly in the injection peak or remain on the column never eluting. Worse, it is difficult to know which is the case for a particular fragment and how to correct the problem.

It is a guessing game to determine the correct eluent concentration – a game the user usually loses unless a new approach is taken.

Isocratic elution for DNA Chromatography must be an organized method that first characterizes the column and eluent through the collection of data, organizes the information into a table and then uses the table to select the conditions needed for the separation. In the work reported here, the method and data are shown for the separation of double-stranded DNA fragments in the 90 – 590 bp range. Similar methods can be used to extend the range in either direction. Also reported are the accuracy and precision of the separation method.

A 3-step-elution process is used to establish the conditions to be used in isocratic elution. The procedure used to characterize the column and eluent strength is shown in Table 6.2. Starting at 46 % B, the injection of a size-based standard is made into an isocratic eluent. The concentration is maintained for 20 minutes and then a shallow gradient is performed reaching a B concentration of 65 over a 15 minutes period. Finally a 100 % B wash is performed and the condition returned to the starting concentration. The key to the process is to use a gradient after the initial isocratic process to elute fragments remaining on the column in a controlled manner. The number of fragments can be counted easily starting from the end of the chromatogram counting backwards so that all of the peaks can be assigned a fragment size. The characterization process is performed in 1 % steps of the isocratic concentration starting at 46 % B and continuing until all of the fragments are eluted in the isocratic region of the chromatogram (about 64 % B).

Table 6.2. Example of elution program for isocratic elution at 46 % B eluent

Isoratic Elution Conditions*

Time (min)	*% A*	*% B*	*Flow Rate (mL/min)*
0.0	54	46	0.75
20.0	54	46	
35.0	39	65	
35.1	0	100	
36.1	0	100	
36.2	54	46	
41.2	54	46	

Solvent A: 0.1 M TEAA
Solvent B: 0.1 M TEAA, 25% ACN
Equilibrium Time: 5 minutes
Acquisition Time: 41.2 minutes

*For 46% B only, other conditions were changed accordingly.

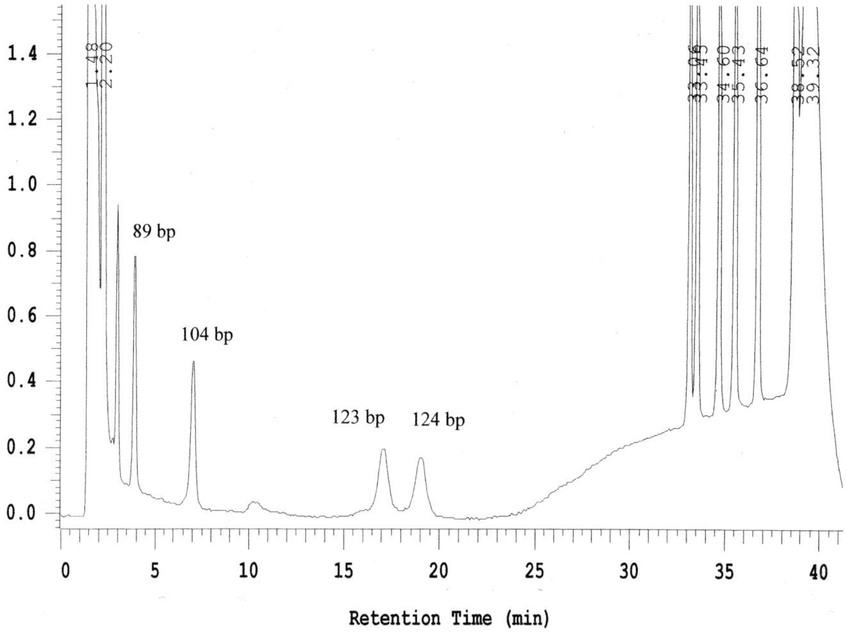

Figure 6.1. Isocratic elution of fragments from pBR322 Hae III plasmid digest using 47 % B eluent (A: 100 mM TEAA, B: 100 mM TEAA, 25 % v/v acetonitrile) at 0.75 mL/min and a preparative DNASep® 7.8 x 50 mm column. The elution program is listed in Table 6.2 except the isocratic eluent concentration is 47 %. A gradient is used at the end of the program to clean the residual material off of the column.

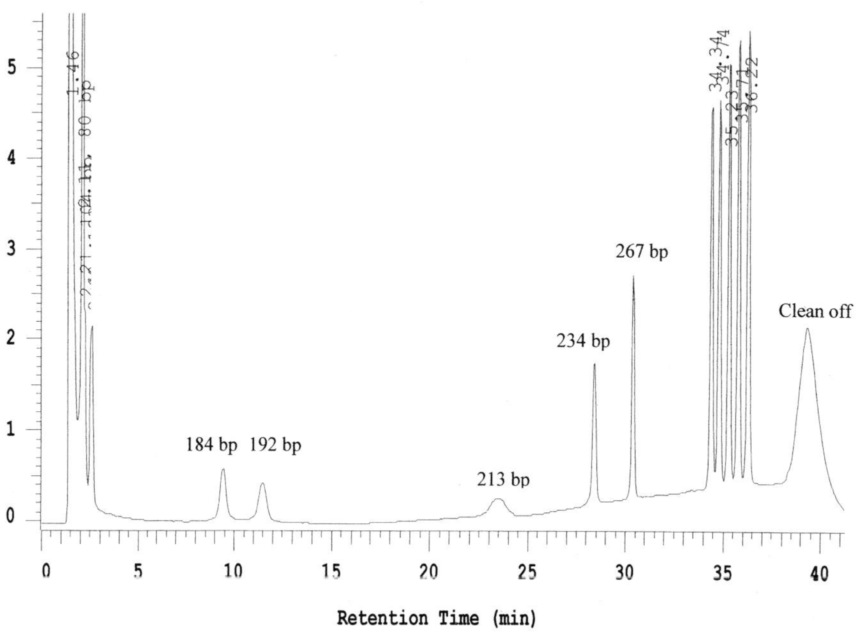

Figure 6.2. Same as in Figure 6.1 except the isocratic elution was at 53 % B.

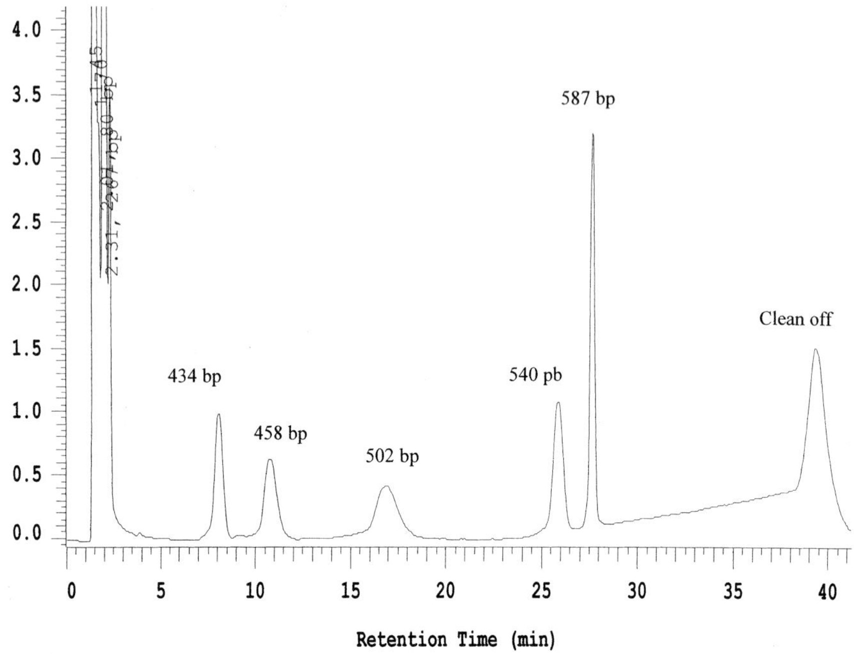

Figure 6.3. Same as in Figure 6.1 except the isocratic elution was at 61 % B.

Figures 6.1, 6.2 and 6.3 illustrate the process. Figure 6.1 shows the separation of 89 bp to 124 bp fragments using a 47 % B isocratic elution. Excellent resolution of the 123 and 124 bp pair has been achieved. Raising the concentration of B to 53 % will separate fragments in the 184 bp to 267 bp range (Figure 6.2). An eluent concentration of 61 % B can be used for fragments in the 434 bp to 587 bp range (Figure 6.3).

The data are collected and assembled into tables as shown by Table 6.3 for pUC18 Hae III and Table 6.4 for pBR322 Hae III. Only one of the standard needs to be used for normal operations. Two standards were used in this work to measure system accuracy (discussed later). Once the data are assembled, determining the conditions needed for the collection of a particular fragment size is a matter of reading the table for the best resolution for the particular size fragment in question. The highest resolution is achieved at the longest retention times. If smaller or larger fragment sizes are of interest, adjustments could be made to the eluent concentration of the elution program to suit the needs of the researcher. Also, the isocratic portion of the elution process could be increased if higher resolution is needed.

In order to test the effectiveness of the procedures an experiment was performed to determine the accuracy of calibration. Table 6.5 shows how the calibration was performed. Fragments from pBR322 Hae III were used as size standards to bracket the size of a fragment of pUC18 chosen to be an "unknown". In this case, frag-

Table 6.3. Retention times of pUC18 *Hae* III fragments under isocratic elution conditions

pUC18 Hae III Digest									
Conc B(%)	***80bp RT (min)***	***102bp RT (min)***	***174bp RT (min)***	***257bp RT (min)***	***267bp RT (min)***	***298bp RT (min)***	***434bp RT (min)***	***458bp RT (min)***	***587bp RT (min)***
46	3.64	9.31	–	–	–	–	–	–	–
47	3.00	6.31	–	–	–	–	–	–	–
48	2.64	4.75	–	–	–	–	–	–	–
49	2.38	3.59	–	–	–	–	–	–	–
50	2.26	3.00	–	–	–	–	–	–	–
51	2.19	2.56	20.64	–	–	–	–	–	–
52	2.13	2.33	11.78	–	–	–	–	–	–
53	2.10	2.19	7.22	–	–	–	–	–	–
54	2.14	2.14	4.89	–	–	–	–	–	–
55	2.08	2.08	3.55	20.01	–	–	–	–	–
56	2.06	2.06	2.82	10.75	13.81	–	–	–	–
57	2.04	2.04	2.41	6.35	7.72	13.78	–	–	–
58	2.01	2.01	2.21	4.36	5.09	8.08	–	–	–
59	2.01	2.01	2.10	3.19	3.56	4.99	–	–	–
60	2.04	2.04	2.04	2.55	3.39	3.41	16.19	–	–
61	2.00	2.00	2.00	2.25	2.31	2.64	8.04	10.71	–
62	2.00	2.00	2.00	2.11	2.11	2.28	4.56	5.62	18.25

ments 267, 243 and 213 were used to draw a calibration curve (Figure 6.4). The data were entered into Microsoft Excel charting and plotting program and an equation describing the line was calculated by the program. The retention time of the "unknown" fragment, pUC18 Hae III 257, was entered into the program and the size calculated. The results, shown in row three of Table 6.6, show that the calculated size of the fragment was 255 bp, 2 bp smaller than the actual size. The experiment was repeated 4 more times as shown in the Table. The left column shows the size range and plasmid used to make the calibration curve. The next column is the unknown fragment size and retention time from the other plasmid. The calibration listed in the last 2 rows used the same plot for two "unknowns". The results show that in every case the calculated size is almost identical or slightly smaller to

Table 6.4. Retention times of pBR322 *Hae* III fragments under isocratic elution conditions

pBR322 Hae III Digest Conc B (%)	80bp RT (min)	89bp RT (min)	104bp RT (min)	123bp RT (min)	124bp RT (min)	184bp RT (min)	192bp RT (min)	213bp RT (min)	234bp RT (min)	267bp RT (min)	434bp RT (min)	458bp RT (min)	502bp RT (min)	540bp RT (min)	587bp RT (min)
46	3.65	5.16	10.60	–	–	–	–	–	–	–	–	–	–	–	–
47	2.99	3.92	7.03	17.12	19.02	–	–	–	–	–	–	–	–	–	–
48	2.62	3.20	5.17	11.05	12.19	–	–	–	–	–	–	–	–	–	–
49	2.42	2.76	3.93	7.20	7.79	–	–	–	–	–	–	–	–	–	–
50	2.25	2.45	3.18	5.20	5.56	–	–	–	–	–	–	–	–	–	–
51	2.12	2.26	2.64	3.73	3.92	–	–	–	–	–	–	–	–	–	–
52	2.12	2.12	2.38	3.01	3.13	16.54	20.83	–	–	–	–	–	–	–	–
53	2.11	2.11	2.21	2.60	2.60	9.45	11.49	–	–	–	–	–	–	–	–
54	2.08	2.08	2.08	2.32	2.32	6.02	7.04	12.71	22.18	–	–	–	–	–	–
55	2.07	2.07	2.07	2.17	2.17	4.16	4.69	7.49	11.80	–	–	–	–	–	–
56	2.06	2.06	2.06	2.09	2.09	3.16	3.45	4.89	6.95	13.86	–	–	–	–	–
57	2.05	2.05	2.05	2.05	2.05	2.58	2.73	3.49	4.54	7.73	–	–	–	–	–
58	2.03	2.03	2.03	2.03	2.03	2.32	2.39	2.81	3.38	5.07	–	–	–	–	–
59	2.01	2.01	2.01	2.01	2.01	2.16	2.16	2.39	2.68	3.56	–	–	–	–	–
60	2.00	2.00	2.00	2.00	2.00	2.06	2.06	2.18	2.31	2.74	16.63	–	–	–	–
61	2.01	2.01	2.01	2.01	2.01	2.01	2.01	2.01	2.01	2.31	8.09	10.81	16.90	–	–
62	2.00	2.00	2.00	2.00	2.00	2.00	2.00	2.00	2.00	2.00	4.58	5.68	7.89	11.63	18.98
63	1.98	1.98	1.98	1.98	1.98	1.98	1.98	1.98	1.98	1.98	3.09	3.55	4.39	5.78	8.29
64	1.97	1.97	1.97	1.97	1.97	1.97	1.97	1.97	1.97	1.97	2.40	2.59	2.92	3.45	4.39

Table 6.5. The 213, 234 and 267 fragments of the pBR322 *Hae III* digest was used to draw a calibration plot (Figure 6.4). In a separate injection, fragment 257 of pBR322 *Hae III* was treated as unknown peak. The calibration plot was used to calculate a size of 255 bp – 2 bp smaller than its actual size (Table 6.6 third row).

pUC18 Hae III Fragment	*G-C Content (%)*	*pBR322 Hae III Fragment*	*G-C Content (%)*
587	43	587	43
		540	56
		502	56
458	43	458	43
434	58	434	58
298	59		
267	51	267	51
257	51		
		234	59
		213	62
		192	47
		184	59
174	59	124	65
		123	56
		104	64
102	55		
		89	61
80	59	80	59

the actual size. The accuracy can be increased by using a calibration standard that is close in size to the sample size being collected.

Accurate collection of material relies on high precision. The separations must be reproducible if they are to be accurate. Making repeated injections of several injections of several fragments tests the precision of the method. The results, shown in Table 6.7, show that the relative standard deviation is less than 1% for the same sample injected 5 times. In a separate experiment, retention times changed approximately 20% over a 5 day period including changing the batch of eluent used. The chart was still usable; no changes would be made to the program for any fragment size. It is likely that a chart could be used for several weeks or

Table 6.6. Four calibration plots and 5 tests of fragment sizes

Fragment Size and Calibration Fragment Sizes (bp)	*Fragment Size (bp)*	*Retention Time (min)*	*Calculated Fragment Size (bp) y=a*e^(bx)*	*a*	*b*
80 – 104 (pUC)	89	5.16	88	0.11062595	-0.043661345
123 – 192 (pBR)	174	11.78	172	0.097788833	-0.027904782
213 – 267 (pBR)	257	10.75	255	0.082891969	-0.0190736
434 – 587 (pUC)	502	7.89	494	0.088684142	-0.009086389
	540	11.63	537		

Elution Conditions used were 46, 52, 56, and 62% B for fragment sizes 80 – 104, 123 – 192, 213 – 267, and 434 – 587 bp respectively.

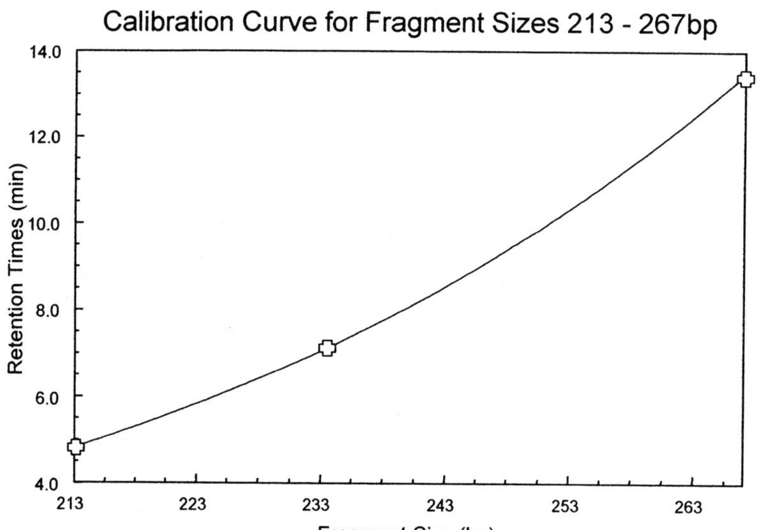

Figure 6.4. Calibration plot of pBR322 Hae III fragments listed in box in Table 6.5 using data from Table 6.4, row 56% B.

even longer if the column is kept clean. Fine calibration for the sample should be performed every day or even with every run depending on the accuracy needed.

Precision also depends on how the injection is performed. Normally TEAA is only added to the eluent and not to the sample. This works well with small (5 µL) injections that are performed on the analytical scale. But purification requires larger injections on preparative columns (e. g. 7.8 mm i. d. column). If insufficient

Table 6.7. Injection Precision Evaluation with pUC18 Hae III Digest

Injection #	*Fragment Size (bp)*	*Retention Times (min)*	*Average RT (min)*	*Standard Deviation (s)*	*Relative Standard Deviation (%)*
1	80	2.05	2.06	0.0122	0.59
2	80	2.06			
3	80	2.08			
4	80	2.06			
5	80	2.05			
1	174	2.74	2.756	0.0207	0.75
2	174	2.75			
3	174	2.79			
4	174	2.76			
5	174	2.74			
1	257	9.97	10.03	0.0696	0.69
2	257	10.06			
3	257	10.13			
4	257	10.03			
5	257	9.96			
1	267	12.80	12.856	0.0611	0.48
2	267	12.91			
3	267	12.93			
4	267	12.84			
5	267	12.80			

TEAA is available to pair with the nucleic acid when it is injected, the peak will front and elute earlier. The retention time and shape of the peak will change with concentration. Adding a concentrated solution of TEAA, so that the sample being injected contains approximately 0.100 M TEAA (just as it is in the eluent), will prevent shifting retention times due to overloading the column. This can be accomplished by using a 2 M concentrate of TEAA and adding 5 % v/v to the sample prior to injection. The capacity of the column will change with sample type and column type. The user is advised to check with the manufacturer.

6.6 Software Collection Methods

Triggering the collection of the fragment is done through the appearance of the peak in the detector. Sometimes a UV detector will be used in series with a fluorescence detector. If two detectors are used in series, then the last detector in the series should be used to determine when the fragment is to be collected.

Detectors can be nondestructive or destructive. UV and fluorescence detectors are nondestructive because material passing through these detectors can be collected and used. A mass spectrometric detector is destructive; but in this case a splitter can be used to divert one stream to the detector and the other to a nondestructive (UV) detector and then to the fragment collector.

Multiple user-selected options are available for sample detection. One of these options is threshold detection that allows collection of all eluent where detector output is above a specified threshold. The collection valve will open when peak absorbance reaches a predetermined level and will switch off when the absorbance falls below that level. Slope detection allows collection of eluent after a rising slope is detected and will stop collection after a falling slope is detected. The steepness of the slope where collection starts and stops is adjustable. Readers who are familiar with HPLC will recognize that the same criteria are used in peak integration for determining peak area.

The most rugged and reliable method of fragment collection is timed collection. Eluent is collected at a predetermined time and collection is stopped at a later predetermined time. The times are chosen based on chromatograms of standards or samples similar to what is being collected. Timed collection can be combined with threshold collection so that only peaks that fall within a specified time period and above a specified concentration will be collected.

The individuals who are undergoing or have undergone treatment for cancer have a need for a sensitive method for detection of the level of cancer. Using classical microscopic methods, the detection of cancer cells in evaluation of a cancer treatment regimen is usually about 1 cancer cell in 100 total cells, or 1 %. Thus cancer cells which are present below the 1 % level will not be seen by traditional analytical methods. A novel method application of fragment collection is called blind collection [9]. Unlike many DHPLC applications where all mutations present are measured in a fragment, this method looks for a particular mutation. It is a method for detection of a putative mutant DNA that includes the steps for amplifying the sample of DNA using PCR. The amplified sample is hybridized to form a mixture of homoduplices and heteroduplices. Using DHPLC, the mixture is separated into fractions by DHPLC with blind collecting the eluted fractions at a retention time corresponding to the retention time of the heteroduplex. The DNA in the blind collected fractions can be PCR amplified to obtain an increased amount of heteroduplex relative to homoduplex. The method is useful for determining the remission status of a patient in which the tissue derived DNA sample contains a large background of wild type or where the putative mutant DNA is below the limits of detection. Much work remains on making this method practical. For example, if there is the problem of PCR induced errors interfering with the actual mutation being measured, the heteroduplex and homoduplex peaks must be highly resolved. Control of these parameters was discussed in Chapter 4.

6.7 Recovery of Material

The need to correctly deposit the material of interest into the vial was discussed earlier. Also, a method for verification of collection was discussed. Once the nucleic acid has been collected, it can be either used directly or concentrated through vacuum evaporation. Triethylammonium acetate (TEAA) ion pairing reagent is volatile and can be removed. Other ion pairing reagents, especially those consisting of quaternary ammonium groups e. g. tetrabutylammonium acetate, are not volatile and cannot be removed with evaporation. In these cases, the nucleic acid can be precipitated with ethanol. Even short, 20-mer single-stranded DNA may be precipitated by the ethanol process.

When evaporating solutions containing TEAA, residual amounts of the ion pairing material may be present when the vial is evaporated to dryness. Ionic neutrality of the nucleic acid must be maintained. The negative phosphate groups on the nucleic acid must have a positive counter ion associated with each group and this most likely would be the triethylammonium cation.

There have been several reports where water was successfully used to redissolve the dried material. However, with the ion pairing reagent present, it is likely that the recovered DNA ion pair complex would seek out a neutral nonpolar surface once it is dissolved in solution. It would seem that taking up the nucleic acid in a buffer would provide a much more stable solution. The following procedures have been successfully used for the recovery of DNA. DNA samples collected from a DNA Chromatogram were lyophilized on a Speedvac instrument (Thermal Savant, Holbrook, NY) for four hours at the medium heat setting. The samples were resuspended in 50 µL of TE buffer [10 mM Tris, 1 mM disodium EDTA, pH 8.0], vortexed, and then used for the appropriate application.

Collection of RNA is performed under the same constraints as DNA as well as several others. Instrumentation and solutions used for RNA must be kept free of enzyme that will degrade RNA. All water used with the instrument and fragment collector is treated with diethyl pyrocarbonate (DEPC). Of course, the instrument should be kept in an area free of dust, laboratory traffic, etc. and gloves should be used for all operations.

RNA samples that are collected cannot be evaporated due to poor stability. RNA samples may be used directly after collection. For example, in some RT-PCR studies, 5 µL aliquots of collected fractions were used directly [10]. Precipitation of RNA is accomplished by taking an aliquot of the collected buffer and adding 10 % (v/v) of a buffer [10 mM Tris-HCL (pH 7.0), 1 mM disodium EDTA, 3.0 M NaCl] and then adding 1 % (v/v) of a glycogen solution (10 mg/mL glycogen) After vortexing the mixture, 2.5 times the original volume of ethanol was added to the solution. The solution was again vortexed. The samples were kept at – 70 °C for 10 min or for –20 °C for two hours before centrifugation at 13,000 g for 15 min at 4 °C. All precipitated RNA fractions were reconstituted in DEPC treated water or buffer [10].

References

1. Transgenomic Application Note 104. PCR amplification and automated sequencing of DNA fragments isolated with the WAVE® nucleic acid fragment analysis system.
2. J. M. Butler, J. M. Devaney, M. A. Marino, and P. M. Vallone, Quality control of PCR primers using in multiplex STR amplification reactions, Forensic Sci. Int. 119, 87, 2001.
3. Transgenomic Application Note 109, Enhancing core facility services with the WAVE® nucleic acid fragment analysis system.
4. Transgenomic Application Note 115. Sequencing of DNA fragments isolated with the WAVE® nucleic acid fragment analysis system.
5. K. H. Hecker, B. Turpie, and A. Kuklin, Optimization of cloning efficacy by pre-coning DNA fragment analysis, BioTechniques, 26, 216, 1999.
6. Transgenomic Application Note 103, Quality control and purification of oligonucleotides and the WAVE® nucleic acid fragment analysis system.
7. M. A. Marino, J. M. Devaney, J. K. Smith, and J. E. Girard, Sequencing using capillary electrophoresis of short tandem repeat alleles separated and purified by high performance liquid chromatography, Electrophoresis, 19, 108, 1988.
8. Transgenomic Application Note 102, Separation and purification of short tandem repeat (STR) DNA fragments using denaturing HPLC (DHPLC).
9. J. L. Sklar, D. T. Gjerde, K. A. Lamb, and C. Hanna, Target DNA amplification by MIPC and PCR, U. S. Pat. 6,218,153, 2001.
10. A. Azarani and K. H. Hecker, RNA analysis by ion-pair reversed phase high performance liquid chromatography, Nucleic Acids Res., 29, e7, 2001.
11. P. D Taylor, R. M. Haefele, D. T. Gjerde and K. A. Lamb, MIPC column cleaning system and process, U. S. Pat. 6,136,195, 2000.
12. D. T. Gjerde and P. D. Taylor, Non-polar media for polynucleotide separations, U. S. Pat. 6,056,877, 2000.
13. D. T. Gjerde, R. M. Haefele, and D. W. Togami, Method for performing polynucleotide separations using liquid chromatography, U. S. Pat. 5,997,742, 1999.
14. D. T. Gjerde, R. M. Haefele, and D. W. Togami, Method for performing polynucleotide separations using liquid chromatography, U. S. Pat. 6,017,457, 2000.
15. D. T. Gjerde, R. M. Haefele, and D. W. Togami, System and method for performing polynucleotide separations using liquid chromatography, U. S. Pat. 5,772,889, 1998.
16. D. T. Gjerde, R. M. Haefele, and D. W. Togami, Apparatus for performing polynucleotide separations using liquid chromatography, U. S. Pat. 6,030,527, 2000.
17. D. T. Gjerde, R. M. Haefele, and D. W. Togami, Process for performing polynucleotide separations, U. S. Pat. 6,156,206, 2000.
18. D. T. Gjerde, R. M. Haefele, and D. W. Togami, Method for performing polynucleotide separations using liquid chromatography, U. S. Pat. 6,174,441, 2001.
19. D. T. Gjerde, R. M. Haefele, and D. W. Togami, System and column for performing polynucleotide separations using liquid chromatography, U. S. Pat. 6,251,272, 2001.

7
RNA Chromatography

7.1
Introduction

RNA molecules play several key roles in the expression of genetic information in all living organisms. The product of transcription is messenger RNA (mRNA). As its name suggests, mRNA carries genetic instructions in the form of a template to the ribosome where protein synthesis takes place. The ribosome is a high molecular weight enzyme comprising ribosomal proteins and ribosomal RNA (rRNA) species, where transfer RNA (tRNA) provides the bridge (anticodon) between mRNA (codon) and the growing polypeptide chain [1]. In addition to these classical roles in cell biology, RNA is emerging as a key regulator of gene expression [2] and has been shown to be capable of catalyzing a range of chemical reactions both *in vivo* and *in vitro* [3].

The double helix is not only the basic molecular form underpinning the human genome; it is an icon of the twentieth century. Research over the last 10 years on DNA repair and recombination have revealed that Watson–Crick base pairing sometimes presents a barrier to many genetic transactions. It is clear that DNA can be transiently unwound, bent, flipped and twisted. But by contrast, RNA molecules are much more structurally diverse. X-ray crystallography and multi-dimensional NMR have revealed the globular nature (to steal a term more frequently associated with proteins) of a number of biological RNA species, and the widely held view that mRNA molecules are just structurally ill defined is proving to be overly simplified.

Our relatively slow accumulation of biochemical knowledge of RNA is largely a consequence of the inadequate methodologies available for RNA isolation, stabilization, purification and characterization relative to its biological partner, DNA. Indeed, while chemical synthesis of 100-mer oligonucleotides is routine, the synthesis of 30-mer oligoribonucleotides is still a synthetic challenge.

The proposal of a *central dogma* of molecular biology by Francis Crick, galvanized molecular biologists into developing experimental approaches for the analysis of RNA. As a result, over the last three decades, RNA has been thoroughly subjected to a battery of centrifugation and electrophoretic techniques. However, our knowledge of RNA structure and function has been dominated until very recently by tRNA biochemistry in a manner similar to the way that haemoglobin and serine proteases have dominated in protein biochemistry over the last 50 years [1].

With the development of methods for the transcription of mg quantities of RNA through the combination of gene cloning and *in vitro* transcription technology, the ability to carry out more rigorous biochemical and biophysical investigations of RNA has arrived. However, there is still a glaring omission in the RNA biochemist's arsenal: methods for high-resolution analytical and preparative separation of the various types of RNA.

One important aspect of translating human genomics into human physiology involves the global analysis of whole cell populations of mRNA in disease and health, and the modulation of these populations during both natural physiological transitions and in response to drug administration. This aspect of RNA population biology is being addressed by micro-arraying technology and to a lesser degree by various RT-PCR approaches including differential message display [4] and serial analysis of gene expression [5]. These forms of analysis of cellular transcriptomes typically produce high volumes of data and treat all RNAs as equivalent molecular species (similar in principle to two dimensional SDS-PAGE (sodium dodecylsulfate polyacrylamide gel electrophoresis) of proteins in proteomics). While in the majority of DNA–DNA hybridization experiments, DNA can be considered as uniform Watson-Crick duplices, the relationship between RNA function and its structure lies somewhere between DNA and proteins. Therefore hybridization-dependent transcriptome approaches provide only a preliminary, and therefore limited, insight into the population biology of RNA.

Here we present a detailed account of the use of chromatography for the separation, stabilization and analysis of RNA. We shall make the assumption here that RNA function is dominated by form. This is readily seen from the classical crystal structures of tRNAs [1]. With these thoughts in mind, at this juncture we shall move on to a discussion of those features of RNA chemistry that may affect its chromatographic behavior.

Since the same chromatographic column and general procedures are used for the separation of RNA molecules as they are for DNA, we shall refer to RNA Chromatography as an analogous process to DNA Chromatography. Still, there is not an all-embracing physical explanation for the separation of RNA species by RNA Chromatography. It is assumed that the mechanism of separation is very similar for RNA and DNA under most conditions, but we shall add the caveat that there may well be different physical principles involved.

RNA is a polymer of nucleotide units, which comprise pyrimidine (cytosine and uracil) and purine (adenine and guanine) bases, the sugar D-ribose and phosphoric acid. The nucleotides in RNA, as in DNA, are combined through phosphodiester linkages between the 5′ hydroxyl of one sugar and the 3′ hydroxyl of another. These polymer chains are not typically found as paired duplices, but rather as complex forms that comprise paired segments and loops that are often stabilized by metal ions. The most familiar structural element in an RNA molecule is probably the stem-loop, in which a non-complementary segment separates two complementary stretches of nucleotides. The resulting structure that often folds is a Watson–Crick-like duplex (the stem) closed off by an unpaired segment (the loop). Such stem loop modules are used routinely to display predicted secondary structures of complex RNA molecules and perhaps the most famous these is the classical cloverleaf structure of tRNAs [1].

The conformation of the duplex segments (stems) in RNA is confined to the A-form. This is due to the steric consequences of the presence of a 2′ hydroxyl group on the ribose ring (DNA, which is predominantly a B-form duplex *in vivo*, has a hydrogen atom at this position) and differences in sugar pucker. As an added structural consequence of the presence of the 2′ hydroxyl group, RNA can engage in an extended repertoire of intra- and inter-molecular hydrogen bonding interactions.

One structural element that demonstrates the clear difference between RNA and DNA is the RNA pseudoknot, in which an anti-parallel complementary duplex arises when the complementary sequences are separated by a sufficient length of chain [6]. The structure takes the form of a handwritten letter f , the upper and lower complementary strands being linked by a pair of 3′ and 5′ loops. This structural diversity, characteristic of biological RNA, has significant consequences for the physical properties of these molecules when attempting to rationalize their chromatographic behaviour. The chromatographic separation of RNA is a function of parameters that include molecular weight (chain length), surface chemistry, morphology and localized single- and double-stranded character. It is likely that unless a given population of RNA molecules is completely freed from secondary and tertiary interactions, by thorough thermal denaturation for example, the chromatography of RNA will be different and unpredictable compared to the chromatography of double-stranded DNA. Even with denaturation conditions, it may be uncertain that RNA has assumed a single-stranded structure if metals are present. However, for a given sample, the separation of RNA is predictable and reproducible. These issues are addressed below along with some practical aspects of RNA extraction and stabilization.

7.2
Biological Extraction of RNA

Cells undergoing RNA extraction (and hands used to do the work) contain a spectrum of ribonuclease enzymes all of which can potentially disturb the original (t_0) population of RNA species isolated from the cell [7]. These enzymes generally catalyze hydrolysis of the phosphodiester linkage on the RNA by deploying suitably ionized histidine residues [7]. Two types of strategies are generally used to overcome the deleterious effects of RNases during RNA preparation. One is the addition of general protein denaturants such as urea and guanidine hydrochloride. Another is the addition of irreversible chemical modification agents such as diethylpyrocarbonate (specific for ionized His side chains). Two key factors to the successful isolation of high quality RNA in good yield are having experience working with RNA and careful laboratory practice.

When RNA from cell extraction is electrophoresed (in agarose), the most abundant species are the two major rRNAs referred to 16S and 23S in *E.coli* (corresponding to 1542 and 2904 nucleotides respectively) or 18S and 28S in most eukaryotes (corresponding to 1874 and 4718 nucleotides respectively). The other major RNA molecules include a species specific tRNA population comprising around 100 molecular species of 60–90 nucleotides in length, a 5S (120 nucleotides) rRNA species (5.8S, 158 nucleotides in mammals) and a small number of spliceosomal RNAs. In contrast, the population of mRNA molecules (the cellular transcriptome) is extremely diverse, ranging from several hundred to tens of thousands of nucleotides in length. Given that there appear to be around 40,000 separate genes in the human genome with the potential for expression of a similar number of mRNAs in any given cell, the population of cellular mRNAs is clearly the most complex of the RNA classes. The protein products derived through translation of the transcriptome determine cellular phenotype. It is because of this large complex population of mRNA molecules that a characteristic background smear is produced underlying the two dominant ribosomal species when a cellular RNA extraction product is submitted to agarose gel electrophoresis and viewed using ethidium bromide fluorescence.

Given the complexity of the RNA population from even the simplest of organisms, it is critical that measures be taken to preserve the integrity of the population prior to separation (or indeed any form of analysis). Since the eluent solvent used in RNA Chromatography is acetonitrile, the RNA must be stable in relatively high concentrations of the solvent and at elevated temperatures (~75 °C). This was shown to be true by simply demonstrating that an RNA species exposed to a high concentration of acetonitrile retained its biological activity. Interestingly, when the effect of acetonitrile upon the activity of a typical ribonuclease enzyme was investigated at a range of solvent concentrations and temperatures, a marked effect upon the potency of the enzyme was observed. At 60 °C in 20 % acetonitrile Bovine RNase is reversibly unfolded reducing the activity of the enzyme. At 60 % acetonitrile at the same temperature the enzyme is also unfolded, but in an irreversible manner [8]. This observation suggests that RNA stabilization, or rather elimination of the deleterious effects of contaminating RNases, can be gained as an added benefit to the separation performed by RNA Chromatography.

In addition to the role of acetonitrile as a solvent in promoting RNA stability through RNase inactivation, the chromatographic process of separating RNA also separates RNase from the RNA fragments. An example of this is Bovine RNase that elutes early in the chromatogram in a separation of RNA fragments. Thus as a consequence of the chromatography, RNA Chromatography will separate (remove) RNase from the large, more complex RNA species that have been recovered from a biological extract.

7.3
Size Based RNA Separation

Figure 7.1 shows the application of RNA Chromatography to separate a series of RNA size standards that are commonly used in agarose gel electrophoresis. At first glance, it appears that under similar denaturing conditions employed for the separation of single-stranded DNA, the RNA species are recovered in a size dependent manner. However as discussed in earlier chapters, the chromatography of single stranded DNA species can produce irregular retention times. These deviations can be even greater for RNA than single-stranded DNA because of the greater variance in sequence of the RNA fragments over most single-strand DNA samples. The additional peak in the chromatogram shown in Figure 7.1 is a 451 nt RNA species that forms part of the human telomerase enzyme. The telomerase RNA elutes between the RNA size standards of 200 and 500 nt, somewhat earlier than would be ideally expected and with an asymmetrical appearance.

A DNA duplex can be considered as a negatively charged cylinder studded with a distribution of hydrophobic patches. Size-based separation of double-stranded DNA is made possible by the structural uniformity and the shielded bases of the DNA double helix. Thus, the suppression of nucleotide sequence-specific hydrophobicity is achieved by forming an ion pair with the phosphate group and an ion pairing reagent to establish the dominance of the hydrophobic (non polar) triethyl moiety. The hydrophobic nature of the resulting molecule is responsible for the principal analyte-stationary phase interaction of the separation.

Ion pairs of single-stranded DNA do not shield the more polar bases so separations of these molecules is not strictly size dependent but also based on polarity of

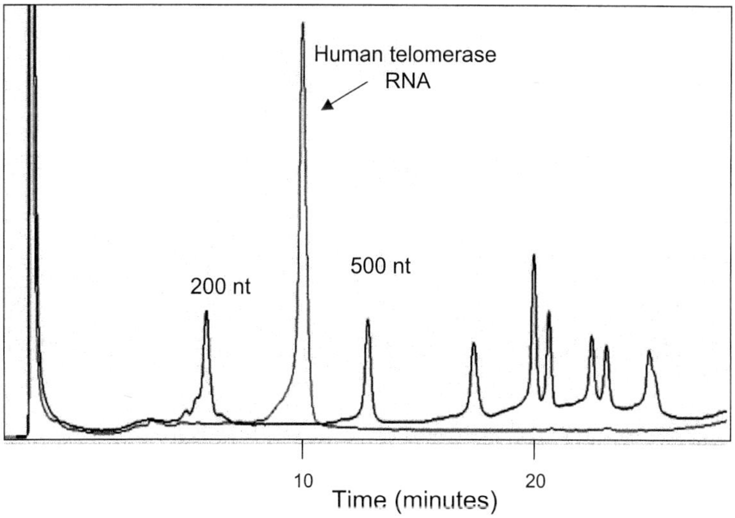

Figure 7.1. Analytical separation of RNA size standards by RNA Chromatography is shown. The 451 nt human telomerase transcript elutes between the 200 and 500 nt species as indicated.

the single strand. Another example of the effect of the single strand nature of a nucleic acid molecule on its retention in DNA Chromatography is shown by the separation of homoduplex from heteroduplex DNA species. The increased single strand character of a heteroduplex leads to a more polar molecule, which in turn reduces the retention time for the heteroduplex. This difference in retention time can be optimized by the manipulation of the temperature of the chromatography run (see Chapter 4 and Appendix 3).

The sequence dependence of single-stranded DNA separations can be reduced by using a more non polar ion pairing reagent such as tetrabutylammmonium bromide and increasing the hydrophobicity of the single-strand DNA/ion pair complex. But in some circumstances, the sequence dependence can prove useful. An example of this is the enhanced retention of T-rich DNA following a bisulphite deamination and PCR in the analysis of DNA methylation (see Chapter 8).

It has been already discussed that RNA Chromatography is best performed at elevated temperatures to denature and reduce secondary of the RNA being separated. However, it is instructive to note that electrophoresis of RNA under denaturing conditions (urea or formamide gels) may not abolish all secondary and tertiary intramolecular interactions. Therefore while some RNAs will be predominantly double stranded, others will be partly folded and others still will be unfolded. In other words, it is impossible to predict what influence will be exerted by temperature and solvent parameters over the chromatographic behaviour of complex RNA populations. Moreover, even solutions of single molecular species should not be assumed to be conformationally homogeneous. In the language of biological NMR spectroscopy, each RNA molecule will exhibit an ensemble of conformations defined by temperature, solvent chemistry, ionic strength and in particular, the presence or absence of divalent cations. Ultimately, any given RNA species like a polypeptide chain will adopt its lowest free energy state(s), a process that is thermodynamically driven by the nucleotide sequence of that particular molecule.

Therefore in considering the mechanism of the separation, the RNA structure must be considered. Due to the single-stranded nature of most RNA species (notwithstanding the formation of intramolecular double strands), there is an unpredictable diversity in structure of fragments in a biological RNA sample, making it very difficult to elaborate any rules for RNA Chromatography.

In the simplest description, there are two modes of RNA Chromatography. One of these modes is equivalent to denaturing DNA Chromatography (akin to the chromatography of single stranded DNA). The second is a specialized form of hydrophobic interaction chromatography. The ion-pairing reagent produces a constellation of hydrophobic patches over the surface of folded and partially folded complex structural forms of RNA.

7.4
Separation of Cellular RNA Species

A typical RNA Chromatography separation of a total RNA extraction is shown in Figure 7.2. There are several distinctive features within the profile (obtained in this case at 75 °C) that are found in most biological preparations of this kind. The earliest eluting species include the population of tRNAs (and probably includes small nuclear RNAs), the middle section of the profile is dominated by the rRNA species and finally underlying the entire chromatogram is a spectrum of mRNAs with many of the fragments centered on the later retention times. Corroboration of the above has been obtained in the case of the tRNA fraction by comparison with commercial tRNA preparations, for mRNA by RT-PCR (see below) and for rRNA, by selectively enriching the polyadenylated mRNA population. Indeed, it is relatively easy to determine the retention time of any RNA species (from an organism whose genome sequence is available) by RT-PCR, followed by nucleotide sequencing.

7.4.1
Separation of Messenger RNA from Ribosomal RNA

The expression of cellular phenotype is a direct result of the protein complement of a cell. The latter is in turn closely, but not inextricably, linked to the population of mRNA molecules produced by selective transcription of the genome in a given cell type. For this reason there has been considerable interest in evaluating both in a qualitative and quantitative manner, the products of transcription. The initial ex-

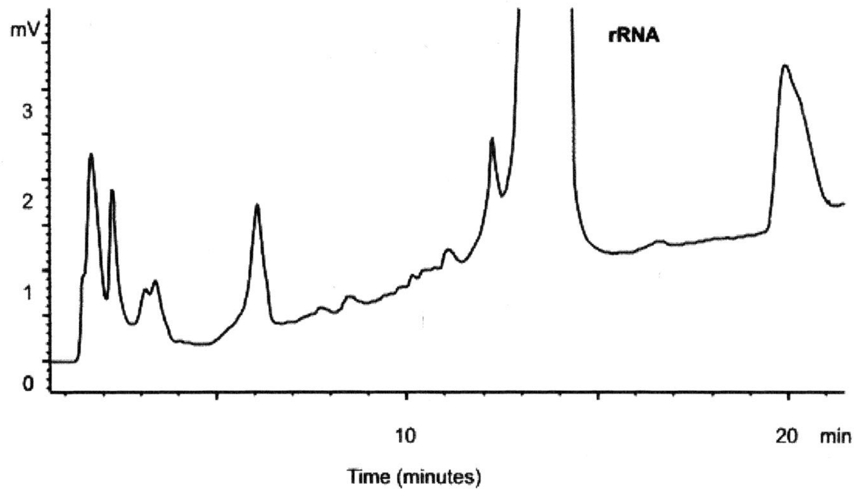

Figure 7.2. RNA Chromatography of a total cellular extract of RNA from tobacco plants is displayed. The large, broad peak eluting between 12–15 minutes is primarily rRNA together with mRNA.

periments in this field were gene specific, but more recently methods have been introduced which facilitate population-based experiments that often utilize microarray technology.

One of the key experimental procedures that forms the preludes to both of the above experiments is the systematic removal of rRNA from the mRNA (polyadeny-

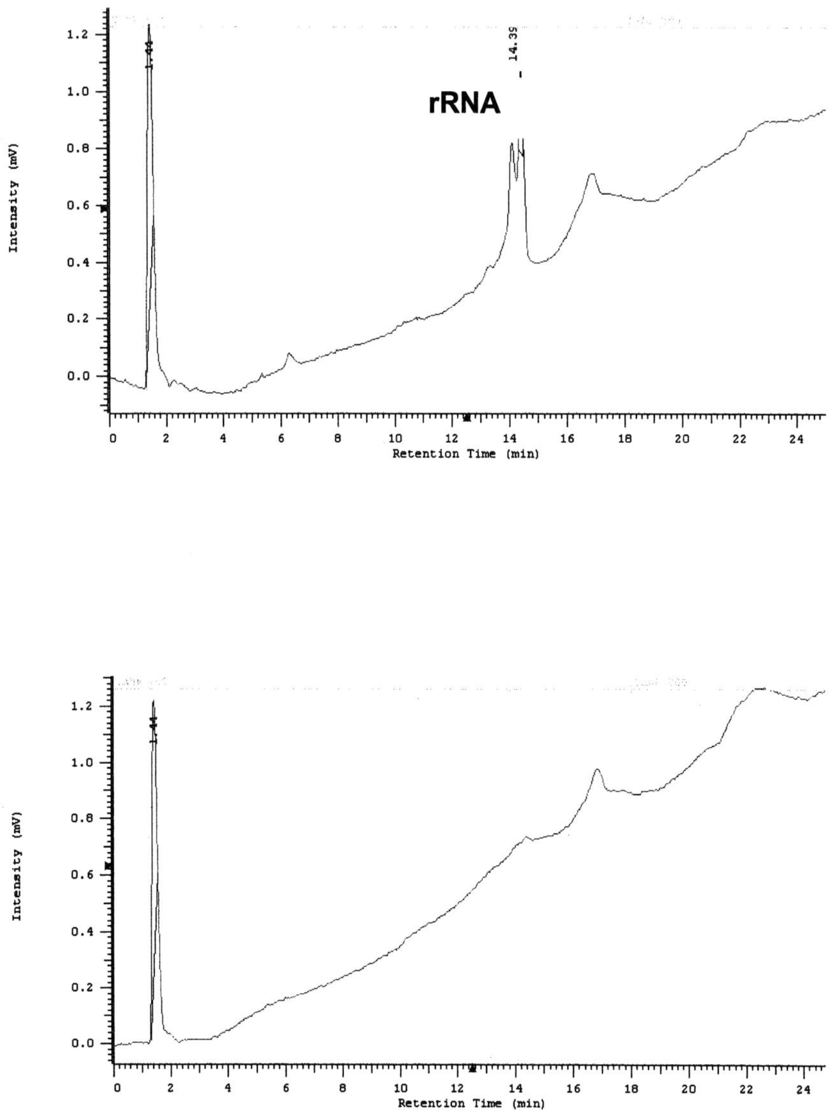

Figure 7.3. Two cycles of polyA enrichment leads to the removal of the major contaminating rRNA species. The upper figure shows the RNA remaining (polyadenylated) after one round of enrichment on an oligodT resin and the lower figure after a second round.

lated) fraction. Typically, the polyadenylated fraction is sequestered via an oligoT chromatography column or batch resin, the rRNA and tRNAs (together with the non-polyadenylated mRNAs) are discarded and, after a suitable washing protocol, the mRNA fraction is concentrated and stored for subsequent experimentation. This procedure also forms the first step in the synthesis of a cDNA library (or simply cDNA) from tissue samples and for the generation of template in a typical RT-PCR experiment.

As can be seen in Figure 7.3, the outcome of a typical mRNA purification can be readily followed by RNA Chromatography. The process serves rather to enrich for mRNA rather than to completely remove the excess rRNA (or rather to deplete the rRNA). Indeed, at least two rounds of enrichment are usually required in order to remove the bulk of the non-polyadenylated RNA. In this particular experiment, RNA Chromatography is used in an analytical mode for the evaluation of polyA-mRNA purification. It is possible to apply the same principle in a preparative mode in order to produce a series of fractions which may facilitate the selective synthesis of cDNA populations for the production of size selected DNA libraries (described in detail in Chapter 8).

One of the major drawbacks of polyA-mRNA isolation is that some mRNAs are not polyadenylated and therefore will be excluded from any subsequent analysis. The use of RNA Chromatography in a preparative mode, offers the potential for isolating at least a fraction of those mRNA species that do not co-elute with rRNA. This is clearly an important area for development, since the analysis of cell specific RNA populations in disease is becoming increasingly important in the molecular medicine.

7.4.2
Analysis of Transfer RNA

Each cell contains a population (usually referred to as a pool) of tRNAs that meet the requirements of that particular cell's (or in the case of bacteria, that organism's) protein synthesis machinery. The range of sizes of tRNA molecules is particularly narrow, compared with mRNA, between around 60–100 nt, and a given cell typically contains around 100 species. While this molecular weight range is ideal for RNA Chromatography, the close similarity of molecular sizes represents a problem for resolving individual components in a typical cellular pool. This is readily seen in Figure 7.4 where the total pool of tRNAs from *Escherichia coli* has been subjected to RNA Chromatography. Nevertheless, it is clearly possible to analyze fluctuations in tRNA pools by RNA Chromatography when used in conjunction with a downstream procedure such as RT-PCR.

The complexity of RNA Chromatography is exemplified by a comparison of a tRNA population and an individual tRNA species. The chromatography of a single bacterial tRNA is shown in Figure 7.5 at two different magnesium concentrations. It is evident that conformational factors dramatically influence the chromatograms. Such factors are likely to be in operation within the tRNA population shown in Figure 7.4. There is a need to conduct a detailed systematic study of the tempera-

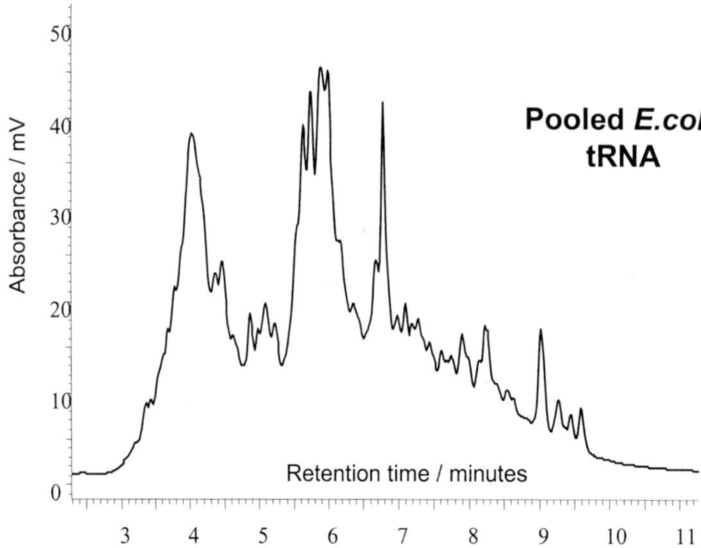

Figure 7.4. RNA Chromatography of the entire pool of tRNAs purified from E.coli. By coupling separation with RT-PCR it is possible to identify and quantify the individual tRNAs. A typical population of tRNAs will contain around 50 species, the abundances of which may be considerably different.

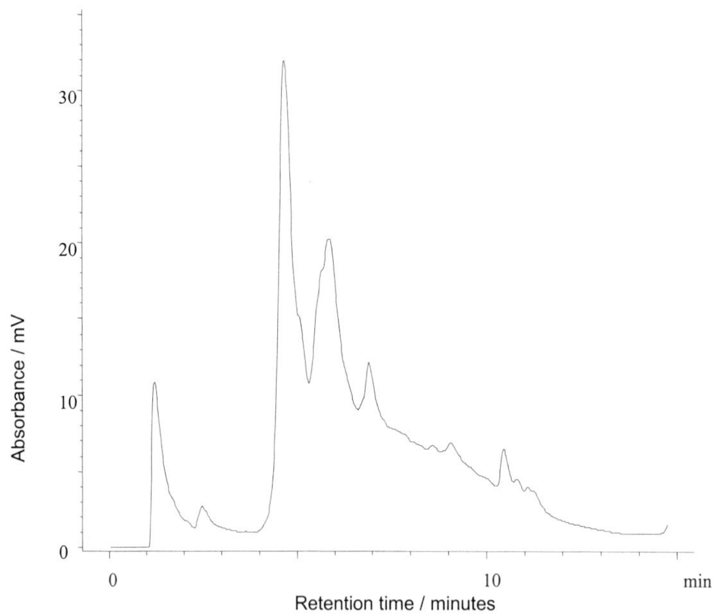

Figure 7.5. Separation of the conformational species arising from pre-equilibration of an individual tRNA with Mg^{2+}. A further dimension of analysis and resolution could be achieved by systematically varying the buffer composition and temperature. In this way RNA Chromatography provide rapid insights into RNA conformation.

ture and solvent influences on the chromatography of complex RNA molecules: the tRNA population represents an ideal test bed for the development of a robust theoretical base for RNA Chromatography.

7.5
Chromatography and Analysis of Synthetic Oligoribonucleotides

One of the major contributors to the rapid rate of progress in contemporary Molecular Biology has been the development of automated synthetic methods for the production of oligodeoxynucleotides. These relatively short (typically between 15–100 nt in length) oligomers are used in the PCR, numerous blotting procedures, transcription factor biochemistry and in some situations have been shown to possess catalytic activity. The development of complementary protocols for RNA synthesis have been slower to emerge, owing to problems arising through the presence of the additional 2′-hydroxyl group. Nevertheless routine synthesis of 20–30-mer oligoribonucleotides is now possible. Such oligomers, as with DNA, have great value in experimental molecular biology, but are of particular importance in the study of catalysis mediated by RNA [3]. Using the hairpin ribozyme [9] as an example, we illustrate below how RNA Chromatography can be used as a general analytical tool in the study of the biochemical and conformational properties of synthetic RNA.

RNA is susceptible to acid-base hydrolysis, which provides a convenient means of optimizing conditions for the separation of oligoribonucleotides at single nucleotide resolution. The chromatogram shown in Figure 7.6 is of a 21 nt synthetic

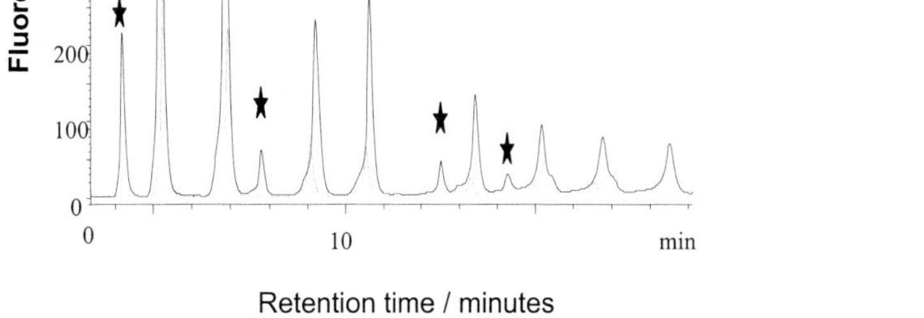

Figure 7.6. The products sequential acid base hydrolysis of a fluorescently end-labelled oligoribonucleotides. During hydrolysis, there is a transient generation of a 2′-3′ cyclic phosphate intermediates (indicated by the starred peaks), this peak is very difficult to observe using electrophoretic separation, but can be readily followed kinetically by RNA Chromatography.

oligoribonucleotides fluorescently labeled at its 5′ end, after exposure to 0.1 M sodium bicarbonate (pH 9 at 95 °C) for 30 minutes. The separation was developed using acetonitrile solvent and tetrabutylammonium bromide ion pairing agent (as described in the figure legend). The high resolution of the smaller cleavage products is apparent.

Acid base catalysed hydrolysis of RNA typically yields two products: a 3′ phosphate moiety and a 5′-3′ cyclic phosphate intermediate which undergoes further nucleophilic attack to yield the 3′-phosphate product. The 3′-5′-cyclic intermediate can be seen using 20 % denaturing polyacrylamide sequencing gels (albeit with some with difficulty). However with RNA Chromatography, these two products are readily identified.

The information encoded by a typical genome is initially transcribed into mRNA and is subsequently translated into functional proteins as discussed earlier in this chapter. However, RNA also has the ability to catalyze biological reactions [3]. This catalytic function is dependent on the 3-dimensional shape of the RNA molecule. One technique that has been employed to obtain RNA structural information is "RNA footprinting". This method provides information on solvent accessibility within RNA molecules and can therefore be used to analyze secondary and tertiary interactions of small RNA structures [10] and RNA–protein interactions [11]. To differentiate between the internal and external regions of the folded RNA molecules, the solvent accessibility of the C5′- and also the generally quoted C4′-position of the ribose moiety can be monitored by the addition of an Fe(II)-EDTA complex together with hydrogen peroxide to the RNA in solution. The hydroxyl radicals generated primarily attack the C5′/C4′-position of the sugar resulting in cleavage of the phosphodiester bond. The cleavage products can then be directly analyzed to identify those sites that show altered solvent accessibility.

Modifications to the standard footprinting reaction [see Chapter 8) that allow the reaction products to be analysed by RNA Chromatography, include the use of fluorescently labeled RNA. This allows the fluorescence-based detection of the cleavage products. The use of tetrabutylammonium bromide as the ion-pairing reagent is essential for the size dependent separation of fluorescently labeled DNA reducing the influence of the hydrophobic fluorescent group and sequence specific effects (see Figure 7.6). Analysis of the footprinting products is rapid with run times of approximately 30 minutes for each sample. Direct quantification of the cleavage products is possible.

In the following example, hydroxyl radical footprinting of the hairpin ribozyme (see Figure 7.7) was performed to analyze the solvent accessibility of the substrate strand as it docks in the ribozyme complex. Figure 7.8 shows the chromatogram generated by base-catalyzed hydrolysis of the fluorescently labeled strand of the ribozyme. The separation of the cleavage products facilitates the alignment of the hydroxyl radical generated cleavage products.

To analyze the solvent accessibility of the substrate strand in the folded ribozyme complex, hydroxyl radical footprinting reactions were carried out on the fluorescently labeled substrate strand. The footprinting experiment was performed in the presence of $Co(NH_3)_6^{3+}$ (Cobalt hexamine is required for folding of the ribo-

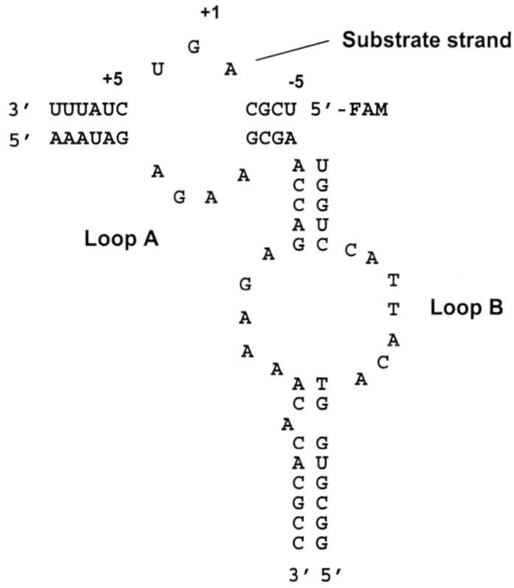

Figure 7.7. This is the proposed secondary structure of the hairpin ribozyme in complex with its substrate.

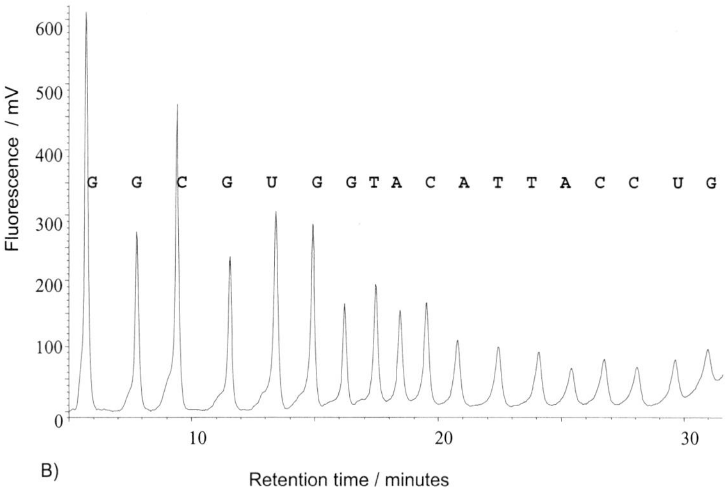

Figure 7.8. This is RNA Chromatography of the synthetic, fluorescent, end-labelled hairpin ribozyme following base-catalysed hydrolysis.

zyme into an active conformation) [12]. The results from hydroxyl radical footprinting of the fluorescently labeled substrate strand in the hairpin ribozyme complex are shown in Figure 7.9. Protection of the substrate was observed in the presence of $Co(NH_3)_6^{3+}$ spanning the substrate cleavage site (A−1, G+1, U+2 and C+3). These results are fully consistent with earlier results [12] that showed the c−2, a−1, g+1 and u+2 are protected, thereby demonstrating that the C5′/C4′-atoms surrounding the cleavage site ribonucleotides are internalized upon folding of

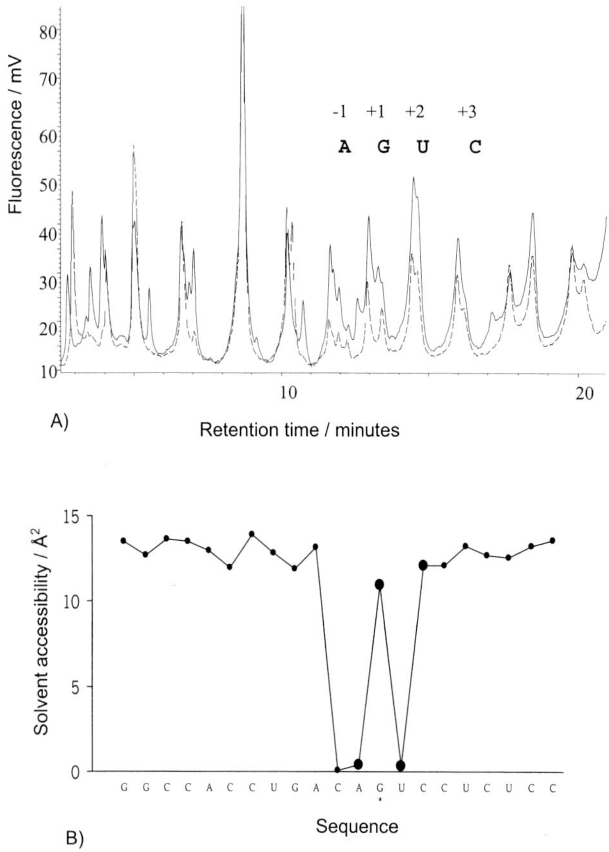

Figure 7.9. The superimposed results of hydroxyl radical cleavage of the hairpin ribozyme are shown in the presence (—) and absence of the substrate strand. The graphical representation below highlights the theoretical solvent accessibility of bases in the complex.

the hairpin ribozyme. These results are also in complete agreement with a tertiary structure model of the hairpin ribozyme proposed in reference [13]. Analysis of the accessibility of the C4′/C5′- positions of the ribonucleotides in the predicted model gave complete agreement with the experimentally observed sites of protection in the substrate strand [14]. No protection of the cleavage products was observed for the fluorescently labeled substrate strand in the absence of loop A and B RNA in the presence of Co^{2+}. Using this novel approach RNA: RNA interactions can for the first time to be analysed in a convenient, quantitative and high-throughput manner.

7.6
Application of RNA and DNA Chromatography in cDNA Library Synthesis

The construction of high quality cDNA libraries is of fundamental importance in contemporary molecular biology, since such libraries play a critical role in the analysis of all aspects of gene expression. Several methods for the construction of cDNA libraries have been described [8]. All of these methods involve a series of enzymatic reactions including first strand synthesis, which is primed by oligo dT and catalyzed by the enzyme reverse transcriptase; second strand synthesis, catalyzed by a second polymerase; end filling, catalyzed by Klenow DNA polymerase, and finally DNA ligation (DNA ligase) into a vector which has often been dephosphorylated [7]. The steps in a typical protocol are shown schematically in Figure 7.10.

A major problem with this multi-step procedure is the frequent failure, or suboptimal yield, of one or more of the steps. In order to ensure that all steps have been successful, careful monitoring at all stages is traditionally carried out by

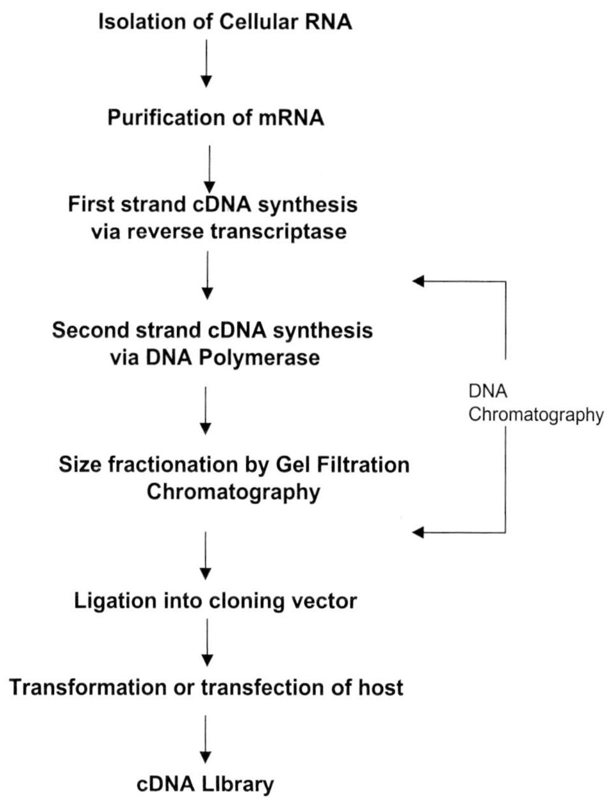

Figure 7.10. These are the typical steps in the preparation of cDNA. The analysis of first strand and second strand synthetic products is greatly improved by DNA Chromatography.

radio-labeling and autoradiography procedures. This normally take one or two days depending on the specific activity of the labeled material. The other problem is that when a pool of fragments, are used for construction of cDNA libraries, the smaller fragments are selectively cloned [7]. There is a need for size-based fractionation of the cDNA especially considering fragment size may differ in molecular weight by an order of magnitude. Sizing is traditionally achieved by gel filtration chromatography and can be very slow. This extended use of radioactive nucleotides increases the probability of radiation damage and requires multi-step analysis. This is usually carried out by gel electrophoresis and an ethidium bromide fluorescence assay in which part of each fraction is lost. Clearly, there are advantages in utilizing the analytical and preparative aspects of both RNA and DNA Chromatography in order to improve these processes.

Analysis of the first strand synthesis reaction (catalyzed by reverse transcriptase) is achieved by subjecting a small fraction of the product to denaturing DNA Chromatography. As shown in Figure 7.11, a series of cDNAs should be produced. The second strand synthesis is usually accomplished through DNA polymerase I catalysis, after which Fluorogreen (Amersham) can be added (at this stage in the protocol synthetic restriction site adapters are typically added to both ends of the double stranded cDNA), and the cDNA fractionated by gravity flow preparative gel filtration. The final products of the protocol can be evaluated both qualitatively and quantitatively by size-based DNA Chromatography (see Figure 7.12).

It is clearly possible in principle to size fractionate the RNA immediately after isolation in a preparative procedure using RNA Chromatography. However, there are issues relating to post-column recovery that must be addressed before prepara-

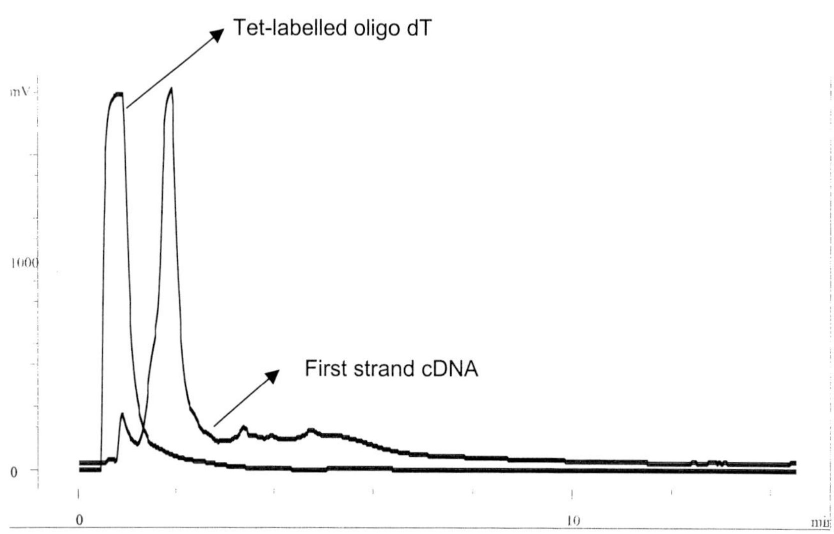

Figure 7.11. The products of a typical first strand cDNA synthesis reaction are separated by DNA Chromatography.

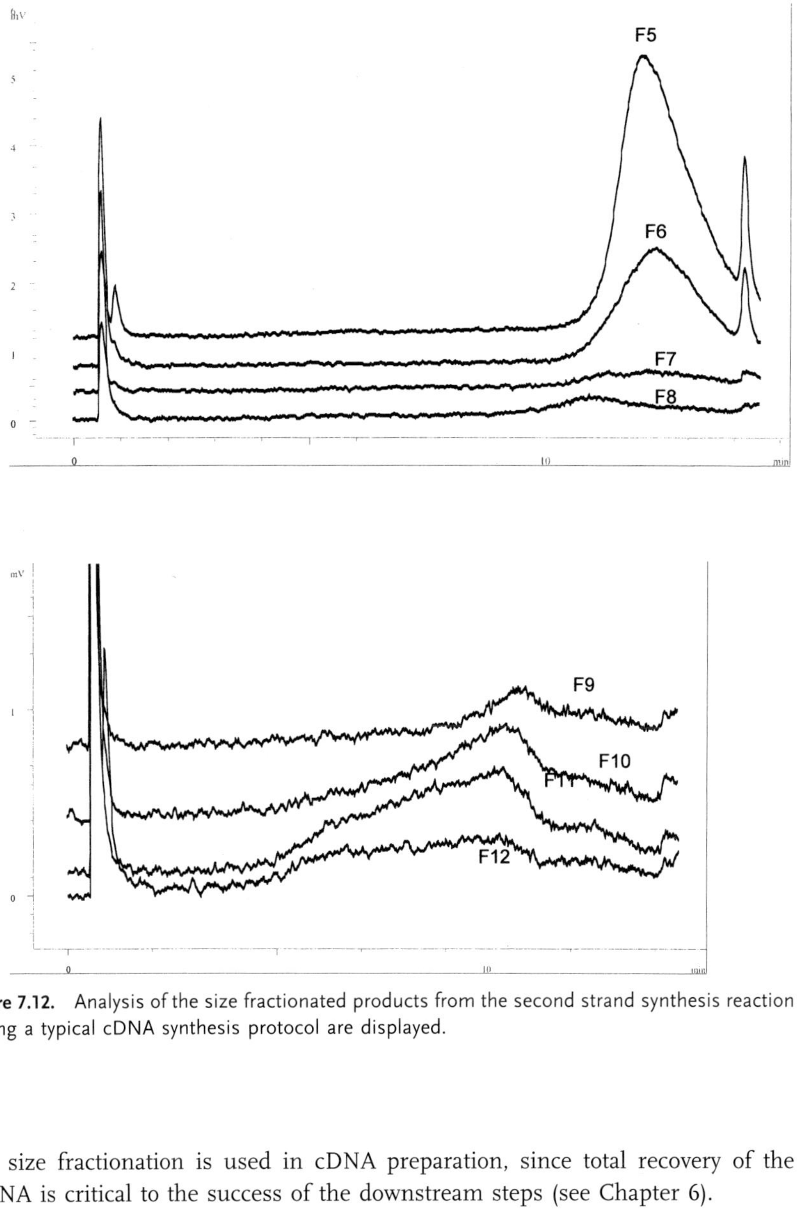

Figure 7.12. Analysis of the size fractionated products from the second strand synthesis reaction during a typical cDNA synthesis protocol are displayed.

tive size fractionation is used in cDNA preparation, since total recovery of the mRNA is critical to the success of the downstream steps (see Chapter 6).

In summary, this method facilitates rapid quality control of cDNA synthesis reactions. It combines the advantages of conventional cDNA library synthesis protocols while eliminating most of their drawbacks. This approach is easy to perform, reliable, quantitative and eliminates the use of radioactivity and therefore makes it a safe alternative to existing protocols. Moreover, there is a potential for preparative RNA fractionation prior to cDNA synthesis in order to further improve the quality of synthetic cDNA libraries enriched for "long" mRNAs.

7.7 Analysis of Gene Expression by RNA and DNA Chromatography

7.7.1 DNA Chromatography Analyses of RT-PCR and Competitive RT-PCR Products

The utility of applying DNA Chromatography to the analysis of RT-PCR products was recognized as early as 1994 [15]. This paper describes the ease-of-use associated with using DNA Chromatography for the analysis of nested RT-PCR products (127 bp and 172 bp) derived from the 5′ non-coding region of hepatitis C virus (HCV) RNA found in human serum. The authors point out that products can be analyzed directly by DNA Chromatography, as opposed to requisite desalting prior to agarose gel analyses. Fluorescein-labeled primers were also used, which enhanced sensitivity 75-fold to give a lower mass detection limit of 2 fmol. This approach allowed for the analysis of 90 samples in 3 hours (compared to 20–40 samples in the same time period for gel-based analyses) and provided more accurate quantification as a result of the technique's chromatographic nature.

The results were also compared to those obtained by ELISA (Enzyme linked immunosorbent assay, the detection of an antigen, usually a macromolecule through recognition of the antigen by a primary antibody, followed by recognition of the specific primary antibody by a "labeled" general secondary antibody). This comparison showed that not all samples that contained antibodies to HCV had detectable serum-borne RNA (i. e. ELISA "false positives"), and antibodies had not developed in some cases of HCV RNA being detectable in the serum (i. e. ELISA "false negatives").

RT-PCR, owing to the intrinsic limitations of RNA stability in tissue samples, and the non-linearity of PCK amplification reactions, is a difficult technique to reproduce accurately. A common means of improving the accuracy of RT-PCR gene expression quantification is to employ an exogenous, homologous "mutant" version of the native target mRNA as an internal standard. This mutant serves as competitor (internal standard) to the target mRNA in both the reverse transcription and PCR processes. The competitor is added to the sample at a known concentration. Assuming the efficiency of each process is the same for the competitor and the native target mRNA, the final ratio of the peak areas of these two RT-PCR products multiplied by the original concentration of the competitor provides a highly accurate means for determining original target mRNA copy number. This general process is typically referred to as "competitive RT-PCR".

A key insight regarding these analyses was provided in 1995 by Doris and coworkers [16] when it was observed that competitive RT-PCR products form stable mutant:target heteroduplices. When gel-based analysis is performed these heteroduplices tend to co-migrate with the mutant product leading to an error in the calculations. This source of error was dismissed in the past as affecting mutant and target bands equally in gel-based analyses and the errors thus "cancel each other out", the investigators showed that uncorrected co-migration leads to large errors when attempting to accurately quantify the level of gene expression.

This problem was overcome by applying DNA Chromatography so that heteroduplices are effectively and quantitatively separated from the mutant and native RT-PCR homoduplices, and these homoduplices are also resolved from one another. Once the three components are resolved and quantified in terms of peak area, it is possible to calculate the *relative* number of final copies of all three duplex fractions since the fragment sizes and molar absorptivities are known. The heteroduplex fraction is proportionally reallocated to the native and mutant fractions so that a simple target-to-competitor ratio is obtained for the purposes of quantification. This provided assays that were highly accurate (95 % recovery for a known amount of starting material), very precise (CVs (coefficient of variation) of 8.3 % to 17.8 % for analyses of rat brain and nephron tissue, respectively), linear over four orders-of-magnitude, and allowed so-called "single-tube" analyses (no titrations with varying concentrations of competitor were necessary). This general approach was applied in subsequent studies for quantification of rat angiotensinogen expression [17], as well as quantification of ER expression in osteoblast cell lines [18]. These last two publications probably represent the most rigorous RT-PCR analyses.

Clearly the single most compelling feature of DNA Chromatography combined with competitive RT-PCR is that of being able to resolve the mutant:target heteroduplices that are likely to form in many instances. However, a key aspect to this approach to gene expression analysis is that one must create and introduce an exogenous competitor mutant mRNA sequence. Furthermore, if the exogenous introduced mutant does not reverse transcribe and/or PCR with efficiencies identical to that of the target, one must determine the exact extent to which these efficiencies differ and apply a suitable correction factor. This is particularly true if one desires to perform "single-tube" assays, whereby perfect linearity is assumed for equal RT *and* PCR efficiencies for both mutant and target. To eliminate the need for prior characterization of RT and PCR efficiencies, investigators examined the effect of mutant homology [19, 20]. It was found that when the mutant was a 145 bp insertion mutant (64 % homology), the efficiency of reverse transcription for mutant vs. target differed by 3.8-fold. However, when the mutant was a 14 bp deletion mutant, the efficiency of reverse transcription for mutant vs. target was identical. Interestingly, the differences in PCR efficiency were the same for the large insertion mutant and the smaller deletion mutant. It was therefore determined that a high degree of homology between mutant and target is needed to achieve good overall accuracy.

One of these studies [20] examined many several of the fundamental features of DNA Chromatography-mediated RT-PCR even further. Again variable co-reverse transcription and co-amplification efficiencies between target and competitor were investigated as a function not only of insertion/deletion mutant length. Also investigated were various reverse transcriptase modifiers; RNA pre-heating procedures (with and without DMSO); concentrations of magnesium chloride, nucleotides, and random hexamers. While changing these conditions led to significant (though small) differences in RT efficiencies, the single largest contributor equalizing the efficiencies between mutant and competitor was the difference in the insertion or deletion. For example, a 24 bp insertion mutant for rat CYP provided identical RT efficiencies as compared to the native sequence. However, a 28 bp deletion mutant for rat could not be equally reverse transcribed as compared to the native sequence. These authors also reported that DNA Chromatography with UV detection, when combined with properly optimized competitive RT-PCR procedures, allows for the reliable detection of as few as 100 mRNA copies.

In an effort to make these analyses simple and available to investigators, a shareware Microsoft Excel Macro (Peter Doris, University of Texas at Houston) was developed for processing the integrated peak data for target, mutant and any associated heteroduplices [21]. The user first prompts to enter information regarding the sizes of competitor and target. Once completed, it is possible to enter data for either titration-based analyses or single-tube analyses (peak areas, competitor mass). The mathematics are performed via the Microsoft Excel macro logic, and results are reported in a spreadsheet format (along with titration plots, if so chosen). A similar program is also available on WAVEMAKER® (Transgenomic, Omaha, NE).

7.7.2 Alternative Splicing

The same authors that pioneered DNA Chromatography-mediated competitive RT-PCR analyses [15], also reported on DNA Chromatography-mediated determination of alternative splicing [22]. In alternative splicing the coding sequence effectively gets larger or smaller by changes occurring in the splicing position, thus providing numerous forms of a protein from a single gene. When expressed, this will have the effect of producing mRNA that carries an "insertion" or "deletion" of a particular length relative to the native target length, which in turn acts as an endogenous RT-PCR "competitor internal standard". This internal standard operates analogous to the exogenous competitors described above for quantification of gene expression. As with quantification of gene expression, accurate alternative splicing determinations require resolution and quantification of native "competitor" heteroduplices to account for all reaction products.

The first paper to describe this approach to detection of alternative splicing describes the biological properties of the 1α and 1β isoforms of RUSH [22]. These investigations required accurate determination of the extent of one isoform's expression relative to the other. DNA Chromatography showed that rush1 is the progesterone-dependent isoform. The authors elaborate upon DNA Chromatography analyses of RUSH alternative splicing in a later paper [23]. They state that the two isoforms differed by a 57-nucleotide insertion (the RUSH 1α RT-PCR product is 225 bp, while RUSH 1β is 282 bp). This insertion introduces a stop codon that prematurely truncates the RUSH 1β isoform. The approach was used for measuring the relative levels of RUSH alternative splicing in a range of tissues.

This approach to detecting alternative splicing has been applied recently by other researchers to thioredoxin-1 (Trx-1) in human cancers [24]. By combining competitive RT-PCR with DNA Chromatography analyses of the reaction products, these investigators demonstrated that alternative splicing might exert some level of control over the amounts of Trx-1 found in certain cancer cells.

7.7.3 Differential Messenger RNA Display via DNA Chromatography

The development of methodologies for the analysis of mRNA populations in individual cells or tissues has revolutionized the investigation of gene regulatory networks in cell biology. Several methods are currently available for rapidly identifying individual and sets of genes that are critical for developmental processes or that mediate cellular responses. Such methods include differential display (DD) [4], comparative expressed sequence tag sequencing [25], representational difference analysis [26, 27], cDNA or oligonucleotide arrays [28, 29] and serial analysis of gene expression [30]. DD has been shown to be a powerful approach in understanding the mechanisms of differentiation and development by detecting altered gene expression in closely related cell lines or tissues [4, 31–34]. The DD technique involves the detection of changes in expression of mRNAs by their selective enrichment without any prior knowledge of the sequence of the specific genes. In a typical DD experiment, total RNA is isolated from those cell types to be compared and first strand cDNAs are synthesized by reverse transcription using an oligo-dT primer that has a specific dinucleotide at its 3′ end. This anchor primer and an arbitrarily chosen primer are then used in the PCR to amplify cDNAs to which both primers can hybridize [4].

The main drawbacks of DD are the lack of reproducibility, the inability to read and compare complex gels and difficulties in recovering the desired bands from gels. Optimized primers and annealing temperatures can reduce the number of false positives [35] and DNA Chromatography offers the potential of overcoming the problems related to comparing and recovering the bands from complex gels.

Experimental data shown in Figure 7.13 from the human embryonal carcinoma cell line NTERA2, a model for human neuronal differentiation [35], was used as a test bed system to establish whether DNA Chromatography could indeed replace the current gel based steps used routinely in DD analysis. Using this chromatographic approach, multiple rounds of amplification are made possible which can be particularly helpful when the amount of isolated RNA is limiting. In addition, preparative chromatography simplifies the recovery and cloning of differentially expressed mRNAs as cDNAs. A schematic representation of chromatographic DD is shown in Figure 7.14.

NTERA 2 cells proved to be an excellent opportunity for validating the chromatographic approach to differential display. Those genes that were shown to be differentially expressed in a preliminary screen are all either genes known to be expressed (or repressed) during retinoic acid stimulation, or carry the hallmarks of those clones of genes that would be likely to be differentially expressed in response to retinoic acid administration. Clearly at this stage, a more extensive analysis of the differentially expressed genes in retinoic acid stimulated NTERA 2 cells is war-

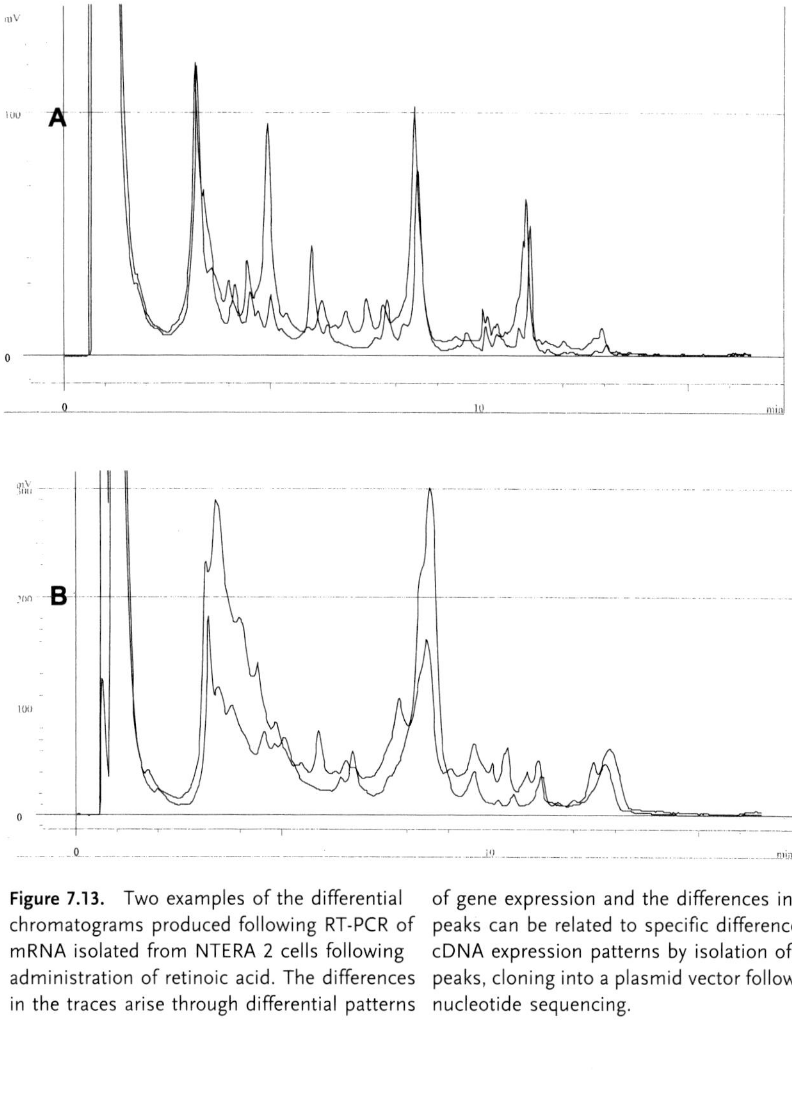

Figure 7.13. Two examples of the differential chromatograms produced following RT-PCR of mRNA isolated from NTERA 2 cells following administration of retinoic acid. The differences in the traces arise through differential patterns of gene expression and the differences in the peaks can be related to specific differences in cDNA expression patterns by isolation of the peaks, cloning into a plasmid vector followed by nucleotide sequencing.

ranted in order to demonstrate the full potential of this method as a tool for gene discovery. These experiments are now underway with a variety of cell and tissue types.

In comparison with high-resolution micro-array analysis in which it is possible to measure the simultaneous fluctuations in expression of thousands of individual mRNA species, chromatographic differential display, as presented here is less comprehensive. Thus individual peaks in a DD chromatogram typically reflect subpopulations of mRNAs. Subsequent rounds of PCR-mediated peak interrogation can approach the resolution (but not the comprehensive coverage) of micro arraying. However, given these limitations, it is the simplicity and low-cost nature

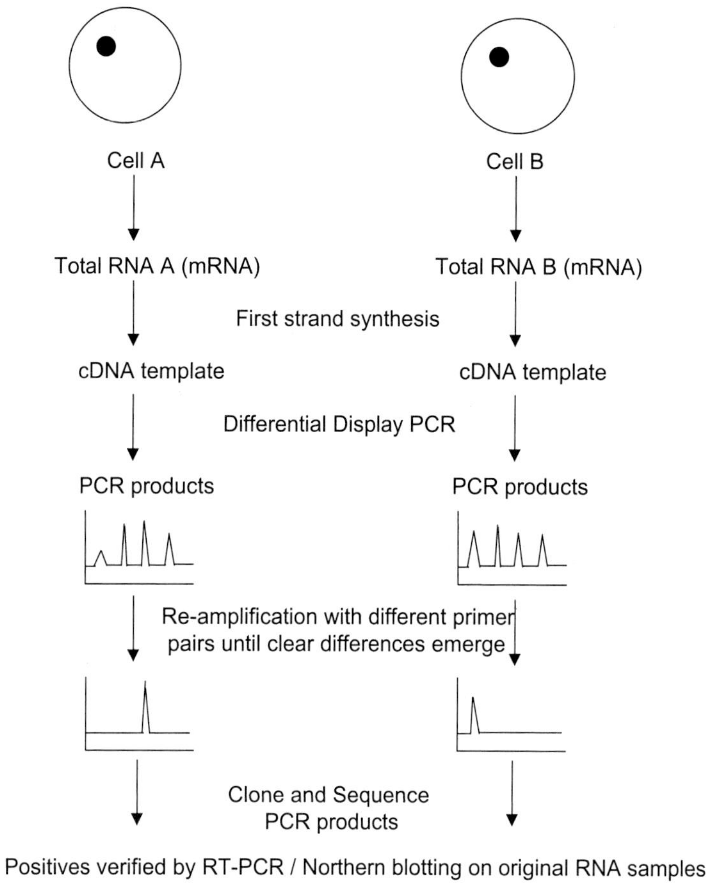

Figure 7.14. Schematic representation of chromatographic differential message display analysis is shown.

of the approach that has been the prime motivation for developing this technique. On the other hand, micro-array analysis can only be applied to the detection of changes in the expression of known genes, and it is often desirable to be able to detect and compare all mRNA species expressed in a particular cell, known and unknown.

References

1. L. Stryer. *Biochemistry*, Fourth edition, W. H. Freeman and Company, New York, 1995.
2. W. Filipowicz, Imprinted expression of small nucleolar RNAs in brain: time for Rnomics, Proc. Natl. Acad. Sci. U S A , 97, 14035, 2000.
3. J. A. Latham and T. R. Cech, Defining the inside and outside of a catalytic RNA molecule, Science, 245, 276, 1989.
4. P. Liang and A. B. Pardee. Differential display of eukaryotic messenger-RNA by means of the polymerase chain-reaction, Science, 257, 967, 1992.
5. V. E. Velculescu, L. Zhang, B. Vogelstein and K. W. Kinzler, Serial analysis of gene expression, Science, 270, 484,1995.
6. J. R. Wyatt, J. D. Puglisi and I. Tinoco, Jr., RNA pseudoknots, stability and loop size requirements, J. Mol. Biol., 214, 455, 1990.
7. J. Sambrook, E. F. Fritsch and T. Maniatis. *Molecular Cloning: A Laboratory Manual,* Second edition. Cold Spring Harbour Laboratory Press, New York, 1989.
8. M. Conroy, J. R. Ashby and D. P. Hornby, unpublished work.
9. M. J. Fedor, Structure and function of the hairpin ribozyme, J. Mol. Biol. 297, 269, 2000.
10. K. J. Hampel, N. G. Walter and J. M. Burke, The solvent-protected core of the hairpin ribozyme-substrate complex, Biochemistry, 37, 14672, 1998.
11. D. Morasand and A. Poterszman, Getting into the major groove Protein-RNA interactions, Curr. Biol., 6, 530, 1996.
12. K. J. Hampel and J. A. Cowan. A unique mechanism for RNA catalysis: the role of metal cofactors in hairpin ribozyme cleavage, Chem. Biol., 4, 513, 1997.
13. P. B. Rupert and A. R. Ferré-D'Amaré, Crystal structure of a hairpin ribozyme-inhibitor complex with implications for catalysis, Nature, 410, 780, 2001.
14. M. J. Dickman, M. Conroy, J. Grasby and D. P. Hornby, RNA footprinting analysis using ion pair reverse phase liquid chromatography RNA, in press 2001.
15. P. J. Oefner, C. G. Huber, E. Puchhammer-Stockl, F. Umlauft , K. Grunewald, G. K. Bonn and C. Kunz. High-performance liquid chromatography for routine analysis of hepatitis C virus cDNA/PCR products. Biotechniques, 16, 898, 1994.
16. A. Hayward-Lester, P. J. Oefner, S. Sabatini, P. A. Doris, Accurate and absolute quantitative measurement of gene expression by single-tube RT-PCR and HPLC, Genome Res., 5, 494, 1995.
17. A. Hayward-Lester, P. J. Oefner and P. A. Doris, Rapid quantification of gene expression by competitive RT-PCR and ion-pair reversed-phase HPLC, Biotechniques, 20, 250, 1996.
18. P. V. Bodine, J. Green, H. A. Harris, R. A. Bhat , G. S. Stein, J. B. Lian and B. S. Komm, Functional properties of a conditionally phenotypic, estrogen-responsive, human osteoblast cell line, J. Cell. Biochem., 65, 368, 1997.
19. P. A. Doris, P. J. Oefner, B. S. Chilton and A. Hayward-Lester, Quantitative analysis of gene expression by ion-pair high-performance liquid chromatography, J. Chromatogr. A, 806, 47, 1998.
20. A. L. Hayward, P. J. Oefner, S. Sabatini, D. B. Kainer, C. A. Hinojos and P. A. Doris, Modeling and analysis of competitive RT-PCR, Nucl. Acids Res., 26, 2511, 1998.
21. P. A. Doris, A. Hayward-Lester and J. K. Hays, Sr., Q-RT-PCR: data analysis software for measurement of gene expression by competitive RT-PCR, CABIOS, 13, 587, 1997.
22. A. Hayward-Lester, A. Hewetson, E. G. Beale, P. J. Oefner, P. A. Doris and B. S. Chilton, Cloning, characterization, and steroid-dependent post transcriptional processing of RUSH-1 alpha and beta, two uteroglobin promoter-binding proteins, Mol. Endocrinol., 10, 1335, 1996.
23. C. A. Robinson, A. Hayward-Lester, A. Hewetson, P. J. Oefner, P. A. Doris and B. S. Chilton, Quantification of alternatively spliced RUSH mRNA isoforms by QRT-PCR and IP-RP-HPLC analysis: a new approach to measuring regulated splicing efficiency, Gene., 198, 1, 1997.
24. M. M. Berggren and G. Powis. Alternative splicing is associated with decreased expression of the redox proto-oncogene thioredoxin-1 in human cancers, Archives of Biochemistry and Biophysics, 389, 144, 2001.

25. M. D. Adams, J. M. Kelley, J. D. Gocayne, M. Dubnick, M. H. Polymeropoulos, H. Xiao, C. R. Merril, A. Wu, B. Olde and R. F. Moreno, Complementary-DNA sequencing – expressed sequence tags and human genome project, Science, 252, 1651, 1991.
26. N. A. Lisitsyn, N. M. Lisitsyn and M. Wigler. Cloning the differences between 2 complex genomes. Science, 259, 946, 1993.
27. M. Hubank and D. G. Schatz. Identifying differences in messenger-RNA expression by representational difference analysis of cDNA, Nucleic Acids Res., 22, 5640, 1994.
28. G. Ramsay. DNA chips: state-of-the art, Nature Biotechnol., 16, 40, 1998.
29. R. J. Lipshutz, S. P. Fodor, T. R. Gingeras and D. J. Lockhart, High-density synthetic oligonucleotide arrays, Nature Genet., 21, 20, 1999.
30. D. Bauer, H. Müller, J. Reich, H. Riedel, V. Ahrenkiel, P. Warthoe and M. Strauss, Identification of differentially expressed messenger-RNA species by an improved display technique (DDRT-PCR), Nucleic Acids Res., 21, 4272, 1993.
31. G. Miele, L. MacRae, D. McBride, J. Manson and M. Clinton, Elimination of false positives generated through PCR re-amplification of differential display cDNA, Biotech., 25, 138, 1998.
32. K. J. Martin and A. B. Pardee, Principles of differential display, Methods Enzymol., 303, 234, 1999.
33. P. W. Andrews, Retinoic acid induces neuronal differentiation of a cloned human embryonal carcinoma cell-line in vitro, Dev. Biol., 103, 285, 1984.
34. K. Malhotra, L. Foltz, W. C. Mahoney, and P. A. Schueler, Interaction and effect of annealing temperature on primers used in differential display RT-PCR, Nucleic Acids Res., 26, 854,1998.
35. M. M. Matin, P. W.Andrews and D. P. Hornby, Manuscript submitted.

8
Special Techniques

8.1
Introduction

Since the introduction of DNA Chromatography for high-resolution separation of DNA, its principal application has been in the arena of mutation detection. Recently, as the acceptance of the resolving power and utility of DNA Chromatography has grown, it has become clear that in principle at least, DNA Chromatography has the potential to replace gel electrophoresis as the method of choice for all aspects of nucleic acid analysis. In this chapter we demonstrate that DNA Chromatography offers considerable advantages over gel electrophoresis in many other areas of molecular biology.

Contemporary molecular biology centers on the metabolism of nucleic acids and has emerged through the cross fertilization of several disciplines including genetics, biochemistry, chemistry and biophysics. In experimental terms, the development of our understanding of nucleic acid synthesis, degradation, modification and expression, has relied heavily upon nucleic acid separation, isolation and detection methods. By far the most widely used method for nucleic acid separation is gel electrophoresis through a matrix of agarose or polyacrylamide. Thus the analysis of the products of a nucleic acid polymerization reaction can be readily observed as the primers and nucleotides become incorporated into larger polymers that can be rapidly resolved on gels. Similarly, the presence of restriction enzyme recognition sites within a given DNA duplex can be readily detected following restriction endonuclease cleavage, by size based separation of the resulting fragments using such gels [1].

In order to illustrate the utility of DNA Chromatography in molecular biology, we shall consider several examples that illustrate how DNA Chromatography can successfully replace gel electrophoresis.

8.2 Analytical and Preparative Enzymatic Cleavage of DNA

One of the fundamental steps in the analysis and characterization of genomic DNA and in molecular cloning is the enzymatic site-specific cleavage and detection of DNA using restriction endonucleases [1]. In addition, as illustrated in Chapters 3 and 6, one of the standard protocols recommended for column calibration and quality control, involves the separation of the products of restriction digestion of a plasmid such as pUC18 with the Type II endonuclease, *Hae* III. Type II restriction endonucleases cleave DNA at specific sites that are usually palindromic, typically containing between 4 and 8 base pairs. The frequency with which a 4 base pair recognition enzyme cuts DNA is approximately once every 250 base pairs (4^4) while a 6 base pair recognizing enzyme cleaves with a much lower frequency of around once every 4,000 base pairs (4^6). Gel electrophoresis would require that both polyacrylamide gels (for the 4 base cutter) and agarose gels (for the 6 base cutter) be used for the separation of these materials. DNA Chromatography can separate these fragments of this size range from both types of cutters using a single column, but with careful choice of the eluent.

An example of the use of DNA Chromatography in the preparative separation of restriction fragments for molecular cloning has been described by Matin and Hornby [2]. When the linear "chromosome" of bacteriophage ΦX174 is hydrolyzed to completion by the 4 base pair cutter *Hae* III, a series of blunt ended DNA duplexes are produced ranging in size from 72 to 1 353 base pairs. Separation of these fragments by DNA Chromatography under non-denaturing conditions yields the chromatogram shown in Figure 8.1a.

A gene " library" can be created if the DNA fragments are pooled, ligated with a suitable plasmid vector and introduced into a bacterial host (typically *Escherichia coli*) by transformation [1]. When this procedure is carried out, it is generally found that the recombinant plasmids recovered from the host will preferentially contain inserts representing the smallest of the DNA fragments of the bacteriophage genome. This phenomenon has a very simple physiological basis: the main-

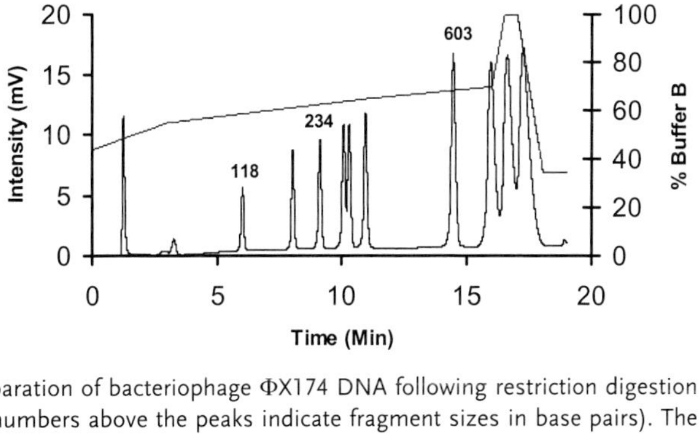

Figure 8.1a. Size based separation of bacteriophage ΦX174 DNA following restriction digestion with *Hae* III is shown (the numbers above the peaks indicate fragment sizes in base pairs). The experimental details can be found in Ref. [2].

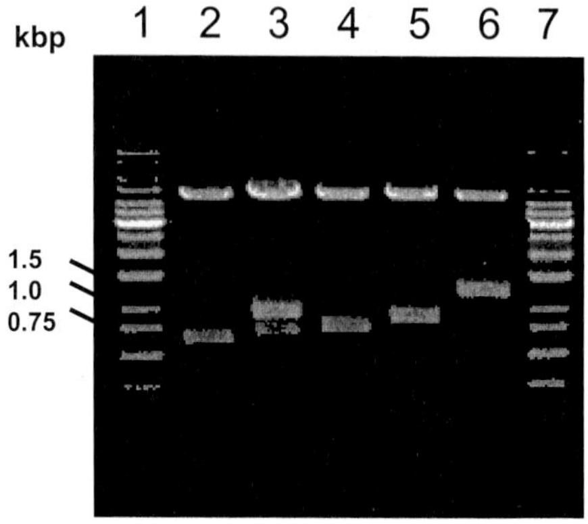

Figure 8.1b. Agarose gel electrophoresis of the recombinant plasmids obtained from the following cloning reactions is shown (described in detail in reference 2). Lanes 1 and 7 are size standards (kbp); lane 2 contains the cloning vector following digestion to highlight the increase in plasmid size upon fragment insertion; lane 3 shows the recombinant plasmids recovered after ligation and transformation of the pooled ΦX174 DNA *Hae* III fragments (Note that only the smaller fragments are recovered as evidenced by the small increase in the size of the molecular weight of the 700 bp vector band shown in lane 1). Lanes 4–7 are the recombinant plasmids recovered after preparative size selection of the ΦX174 DNA *Hae* III fragments prior to the ligation reaction.

tenance and replication of plasmids harboring large inserts places greater energy demands on the host organism than plasmids containing small inserts.

By size fractionating the genomic restriction fragments in a preparative mode using DNA Chromatography, it is possible to "control" the range of insert sizes produced in such cloning experiments. Thus in Figure 8.1b, if the peak fractions recovered from chromatography provide the input for the ligation reaction, all recombinant plasmids recovered following transformation, carry the predicted insert fragments. Indeed in the method described by Matin and Hornby [2], by deploying a positive selection vector, only recombinant plasmids of the correct structure are recovered. The coupling of preparative chromatography and restriction endonuclease digests of DNA will greatly facilitate the generation of size-fractionated gene libraries [2].

The process, mechanism, rate expression and rate constants of enzymatic action on a fragment can also be measured by DNA Chromatography. An enzyme can be added to a vial (reaction vessel) containing a known plasmid or fragment. The reaction conditions of the enzyme can be studied by controlling the temperature, reactant concentrations, etc. Periodically, an aliquot is taken from the reaction vial and injected into the instrument. This in effect freezes the reaction and the concentrations of the reactant and product components can be measured. The enzyme stops reacting with the nucleic acid because it is separated from the sample on the column and a "snap shot" of the process can be taken. The various fragments will increase and decrease in concentration as the action of the enzyme continues until at the end of the reaction the distribution of the fragments is at its final form. If the reactant is a plasmid, the final molar concentration of the various fragments is equal (one of each fragment type is produced from each plasmid). The mass concentration of each fragment depends on the relative size of each fragment produced. This experimental approach has been widely used in the study of mechanistic diversity amongst restriction endonucleases [3], but has until now relied on electrophoretic separation and the use of tritiated plasmid DNA to provide accurate quantitative data: such experiments are greatly facilitated by DNA Chromatography.

An example of the analysis of an endonuclease catalyzed reaction is shown in Figure 8.2, where cleavage of a DNA heteroduplex by the mismatch endonuclease CEL I [4] has been analyzed as described above. Using two fluorescent oligodeoxynucleotide duplices containing defined mismatches together with a homoduplex control, the products of the reaction are revealed under denaturing conditions by DNA Chromatography. The appearance of a set of products differing by one base is indicative of relaxed cleavage specificity or intrinsic exonuclease activity associated with this enzyme.

There is an interesting side note to this work. As stated, the cleavage process depends on first attachment of the enzyme to a specific sequence within the fragment. It is possible with DNA Chromatography to study this process. Attachment of an enzyme to a fragment may alter the hydrophobicity of the fragment and therefore increases the retention time. If the attachment is stable, then the fragment/enzyme combination can be separated. A chromatogram of this type was shown [5] where *Hae* III was reacted with a 500 bp DNA fragment with 4 cleavage

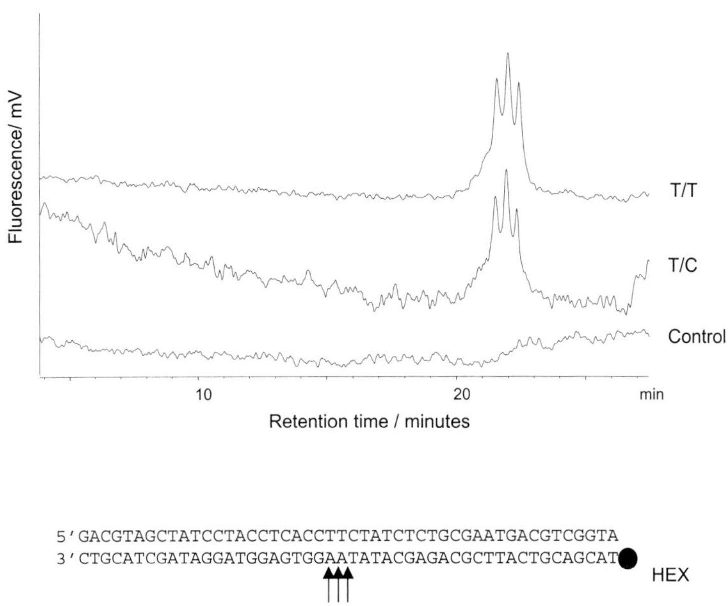

Figure 8.2. CEL I catalyzed cleavage of two heteroduplexes compared with the refractory nature of a homoduplex control molecule. The control duplex Is a homoduplex of the sequence indicated carrying a fluorescent label on one strand. The two heteroduplices are indicated by T/T and T/C. The chromatogram was developed under fully denaturing conditions as described in [13]. Note the localized cleavage adjacent to the mismatch as shown by the split appearance of the cleavage peak.

sites (2 at each end of the fragment). After 2 minutes of reaction an injection was made and the enzyme/fragment complex was measured. After further reaction, all the sites were cleaved and the final products measured. Gjerde and Taylor [5] also discuss the use of enzymes to detect and selectivity nick or cleave mutation sites as a method of determining the presence of a mutation. The use of DNA Chromatography to study enzymatic action has only been performed on a limited scale. But given the ability to control and stop the reaction and to measure products and reactants rapidly and accurately, the technique is certain to be used extensively in the future by nucleic acid biochemists.

8.3 Analysis of DNA Methylation

Following biological replication, DNA may become modified by methylation, extended by the action of telomerase, damaged by alkylation, interrupted by recombination or mutated by erroneous repair [6]. This intrinsic plasticity of the genome contributes to the phenotype of a given cell or tissue and paves the way for evolutionary change. Therefore, there has been substantial interest shown in the molecular processes that lead to these and other DNA modifications. For example, the phenomenon of genomic imprinting and some aspects of developmental gene regulation are intimately associated with region-specific cytosine methylation [7]. In some circumstances, methylated DNA can be distinguished from unmethylated DNA by the protection that methylation affords against endonuclease attack at certain sites. The comparative use of *Hpa* II and *Msp* I has proved to be highly informative in defining the methylation status of those CpG sites embedded in CCGG sequences. Hpa II cannot cleave CCGG sites if the central CpG is methylated. *Msp* I will cleave a methylation site at the internal C, but interestingly cannot cleave the methylated 5′ C site.

In a typical experiment for analyzing the tissue specific methylation status of a given gene, genomic DNA samples are exposed to *Hpa* II and *Msp* I separately and the products of digestion are separated by electrophoresis. Following Southern Blotting, a gene specific (usually radioactive) probe is used to compare the susceptibility of a specific region of the genome to methylation: methylated genes appear as larger fragments on the Southern Blot autoradiograph compared with those that are methylation free.

In a lower resolution alternative to this approach, it is often useful to obtain the bulk differences in methylation of two genomic DNA samples e. g. assessing the global methylation shifts during embryonic development. In this case, the yield (and sometimes the nucleotide sequences) of *Hpa* II fragments is obtained for the two comparative genomic DNA samples [8]. Once again radioactive labeling (in this case end labeling of the products) is necessary in order to compare the yield of the small *Hpa* II fragments; fewer fragments and a different series of sequences are produced in the sample of genomic DNA with the greatest level of methylation [8]. An example of this approach is illustrated in Figure 8.3

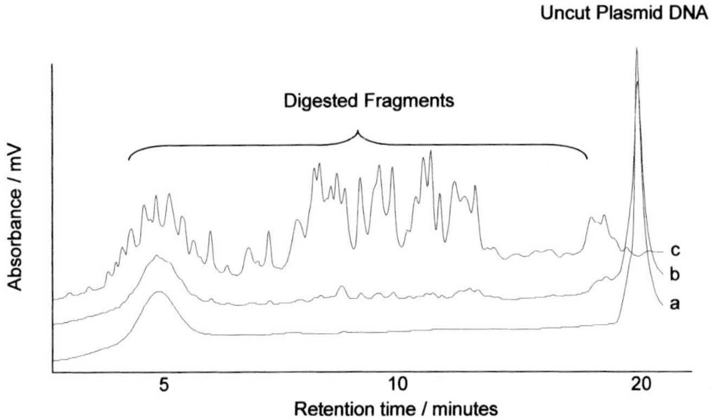

Figure 8.3. The analysis of (plasmid) DNA methylation status by differential susceptibility to *Hae* III digestion and DNA Chromatography. Three plasmids which are fully methylated (trace a), partially methylated (b) or unmethylated (c) were exposed to complete digestion with *Hae* III restriction endonuclease and the products separated under size based conditions by DNA Chromatography. The differences in the chromatograms are due to the refractory nature of the methylated plasmid (which produces a single late eluting peak in a) and the differential levels of cleavage products obtained in traces b and c. The differential susceptibility of methylated and unmethylated DNA is a at the heart of the biological phenomenon of restriction and modification and this has been successfully developed as an in vitro analytical technique.

where the susceptibility of three differentially methylated plasmids to cleavage by *Hae* III can be readily detected by DNA Chromatography. However, it is not possible to replace the electrophoresis and blotting approach to one involving DNA Chromatography when genomic DNA is the subject of the investigation. Moreover, the sensitivity of these analytical methods is inadequate in those situations where the CpG sequence under examination is not part of a convenient restriction site, for which a pair of isoschizomers (restriction endonucleases that cleave the same DNA sequence) showing differential sensitivity to CpG methylation exists.

Higher resolution methods have recently been introduced in order to identify those specific cytosines that undergo methylation and demethylation [9]. Such methods are generally based on the differential susceptibility of methylated and unmethylated cytosines to sodium bisulfite mediated deamination [shown schematically in Figure 8.4]. Following differential deamination, those bases that were originally methylated remain as Cs and those that were unmethylated become Us. In the final stage of analysis, which typically involves PCR amplification, methylated C:G base pairs are retained in the PCR product, while unmethylated C:G base pairs are converted to T:A, and nucleotide sequence analysis provides a facile method of distinguishing the two.

Although deamination of cytosine proceeds at a significantly greater rate than deamination of the corresponding methylated form [9], the reproducibility of the method is often unsatisfactory. Still, with proper controls, bisulfite sequencing has become the accepted approach to the robust analysis of the methylation status

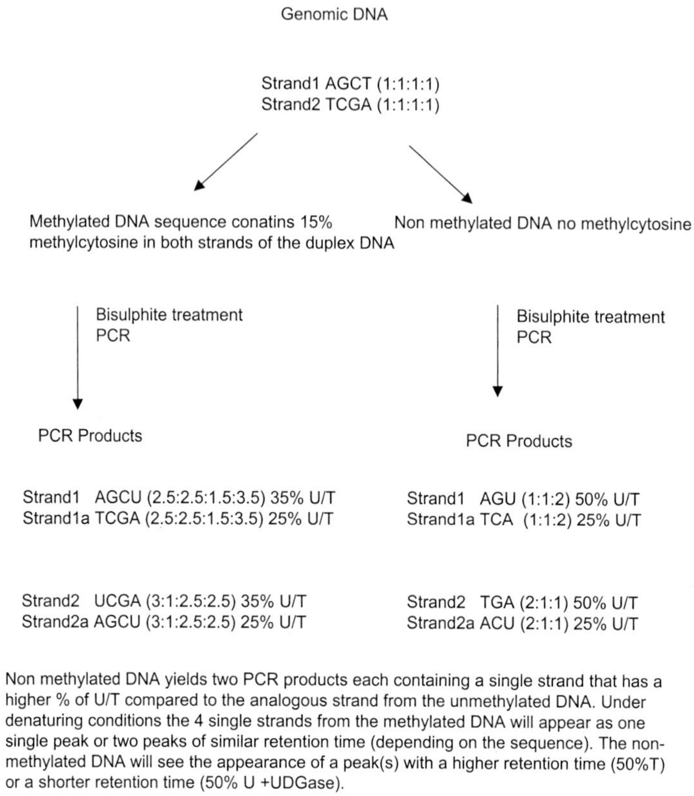

Figure 8.4. An example of bisulfite mediated DNA methylation detection. The principle of the method lies in the differences in retention time between uracil containing versus thymine containing DNA strands that arise through the selective deamination of unmethylated cytosines (rather than those modified at the C-5 position) in a DNA. Additional indirect information about the methylation status of the original DNA can be obtained by treatment with UDGase as described in Section 8.4.3.

of a given genomic segment. DNA Chromatography can provide a rapid means of evaluating the outcome of a deamination reaction. Thus in Figure 8.5, the deamination of a specific cytosine in the oligodeoxynucleotide homoduplex containing the sequence GAATTC (an *Eco*R I restriction site) can be assessed by the outcome of an *Eco*R I restriction endonuclease reaction as a function of the rate of a bisulfite deamination reaction. If bisulfite deamination is 100 % successful, there will be no cleavage of the site since the sequence GAATTU is not a substrate for the enzyme; incomplete deamination is manifested by the appearance of two short duplices. Figure 8.5 shows the successful outcome of a deamination reaction together with those species expected if deamination is incomplete. This quality control experiment, which is much easier to quantify by chromatography than electrophoresis, can lead to a significant improvement in the reproducibility of bisulfite reactions when carried out in parallel with deamination reactions on PCR products.

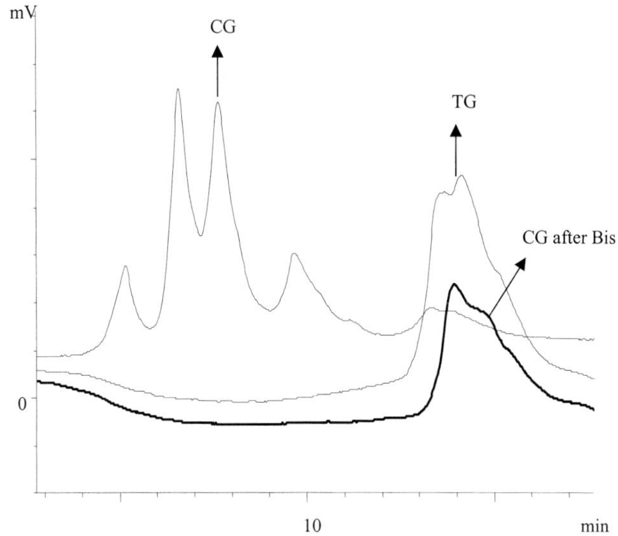

Figure 8.5. Indirect monitoring of busulfite mediated deamination of cytosine containing DNA duplices. An oligodeoxynucleotide duplex containing a unique, asymmetrically positioned *Eco*R I restriction site (GAATTC), was digested with an excess of *Eco*R I before (labelled CG) and after (CG after Bis) bisulfite deamination. As a control, a duplex in which the *Eco*R I site was mutated to GAATTT (labelled TG) to mimic deamination, was also subjected to denaturing DNA Chromatography. By linking restriction endonuclease susceptibility to deamination, this set of DNA duplices can be used to evaluate the outcome of any deamination reaction in a quality control manner. The deaminated strand is clearly distinguished from the cleaved species.

This can be important given that the variation in genomic DNA preparations and in the quality of the bisulfite reagents mitigate against reproducibility. DNA Chromatography in conjunction with bisulfite deamination is well suited to the investigation of DNA methylation. Several approaches are discussed below in which chromatography replaces gel electrophoresis.

Methylation Specific PCR (MSP) is a relatively recent technique that facilitates the precise mapping of DNA methylation patterns in CpG islands [10]. This procedure again takes advantage of the bisulfite-mediated chemical conversion of cytosine to uracil, followed by PCR, using primers designed to distinguish methylated from unmethylated DNA. The ability to amplify identical DNA sequences that differ only in methylation status forms the basis of MSP. Primer design is a critical and complex component of the procedure. Primers must be designed so that mismatches are created which prevent mis-priming between the primer sets and the undesired target DNA. A typical experiment will involve performing two PCR reactions using the same template DNA.

One reaction uses primers (the "U" primer set) designed to anneal to the template if it is unmethylated, and the other reaction utilizes primers (an "M" primer set), designed to anneal if the template was originally methylated. If the sample DNA was originally unmethylated prior to chemical modification, only the "U" primer set will produce an amplification product. Conversely, a product will be produced only with the "M" primer set if the DNA was originally methylated.

In the method described here, initial bisulfite treatment of the DNA is also used, however amplification products will be generated if the template is methylated or unmethylated. The methylated DNA will produce amplification products which posses single strands of DNA with a lower proportion of thymine compared to the unmethylated strand (a hypothetical case study is shown in Figure 8.6). It is this percentage difference in Ts that can be exploited to determine methylation status. Thymine is the most hydrophobic of the naturally occurring bases and as such can give rise to a sequence dependent effect on the separation of oligodeoxynucleotides using DNA Chromatography. Therefore in principle the separation of the single strands (of duplex DNA) of the same size under denaturing conditions should be possible if the proportion of T between the two single strands is sufficiently different to confer additional hydrophobicity in one strand over the other.

To demonstrate that separation of single strands from duplex DNA PCR products is possible based on sequence dependent hydrophobicity differences, a 227 bp PCR product in which the proportion of Ts in one strand was significantly different to the complementary strand was analyzed. The percentage difference in T between the two strands was 22%. Figure 8.6 shows the 227 bp PCR products

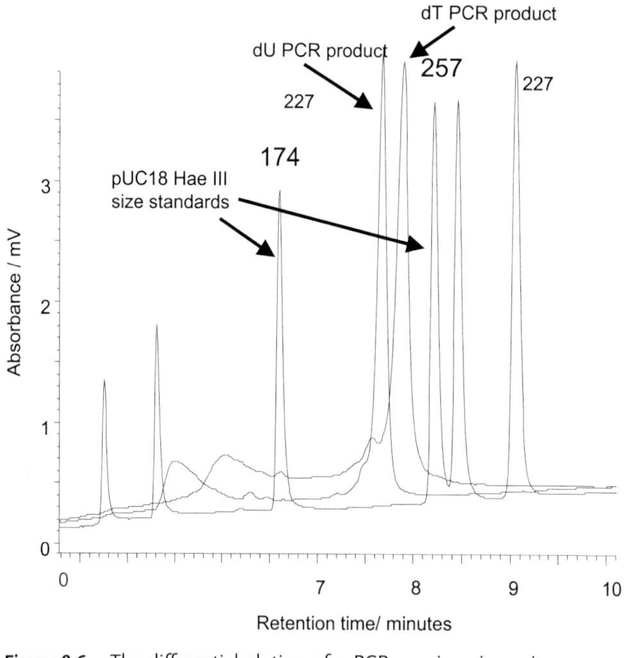

Figure 8.6. The differential elution of a PCR product containing dU versus one containing T. Under conditions of size based separation, an all U duplex (which would arise when the cytosines in a given sequence are unmethylated) elutes earlier than its T counterpart. This difference in hydrophobicity between dU and T is likely to be mimicked by dC and 5methyl-dC.

run in comparison with the pUC18 *Hae* III size standards. The results show the clear difference in retention time of the PCR products with dTTP and dUTP. The uracil containing duplex DNA runs with a significantly shorter retention time compared to the thymine containing duplex, demonstrating that the differences in hydrophobicity are sufficient to alter retention time.

An alternative technique using the same principle involves amplification of the differentially methylated DNA using dUTP instead of dTTP in the PCR. This will produce amplification products with a higher proportion of uracil from the unmethylated DNA. Subsequent treatment with uracil DNA glycosylase (UDGase: see 8.4.3) will remove any uracil bases from DNA, thereby reducing the net strand hydrophobicity. Therefore, the hydrophobicity of the single strand that contains a higher fraction of U will be reduced by a larger extent compared to the strands that contain a lower fraction of uracil (methylated DNA). This difference in hydrophobicity can again be exploited to separate the single strands, which differ in their uracil content (an example of the use of UDGase to alter the hydrophobicity of the DNA is shown in Figure 8.7). The separation achieved using this method was superior to that based on thymine difference alone, however, lower u. v. absorbance (owing to a net loss of nucleotide bases) and small amounts of DNA degradation (owing to abasic site formation) is often observed using this procedure.

The principle of sequence dependent separation of single stranded oligodeoxynucleotides under denaturing conditions can therefore be applied to the detection of DNA methylation differences. It is important to note that primers that bind to predominantly ATG sequences in the target sequence must be used, so that amplification products are generated from both bisulfite treated and non bisulfite-treated DNA and will bind to both methylated and non-methylated DNA.

Since it is relatively simple to resolve DNA duplexes of, for example, 24 and 25 bps, by combining primer extension with bisulfite deamination, it is possible

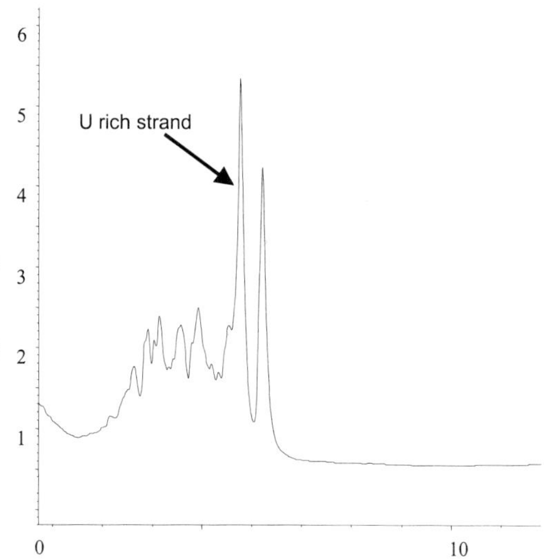

Figure 8.7. The separation of a U rich strand can be enhanced by selective removal of the base via the reaction catalyzed on both single-stranded and duplex DNA by uracil glycosylase (UDGase). The U rich strand indicated in the Figure elutes earlier following base removal, which reduces the net hydrophobicity of the DNA strand during DNA Chromatography under denaturing conditions.

to use DNA Chromatography for the detection of imprinting disorders such as Prader-Willi and Angelman syndromes [11]. Following cytosine deamination of the test and control genomic DNA sample (and as a control for the deamination reaction, two differentially methylated plasmid DNA templates should be included). The resultant PCR products were then used as templates for primer extension reactions, and the products were analyzed using DNA Chromatography. Primer extension reactions were carried out following published methods [12]. The success of such experiments is critical on the complementarity of a primer sequences to a region that is the same whether methylated or unmethylated following bisulfite treatment. By using the appropriate deoxy- and dideoxynucleotides and sizing the products it is possible to distinguish between the presence of a C or T, which defines methylated or unmethylated cytosines in the starting DNA. For example, if a 20 bp forward primer is synthesized complementary to a sequence 5′ to a C̲GT sequence of a normal sample, by using ddTTP in the reaction, the extension product will be 23 bps if the C is methylated, and 21 bps if it is not methylated. The results of such analyses are shown in Figure 8.8.

This method has several advantages over the methods described earlier since it circumvents the use of restriction enzymes (which are generally limited to the analysis of CCGG sequences) and the use of radioactive nucleotides is unnecessary. Moreover, this method is not based on the detection of simply the presence or ab-

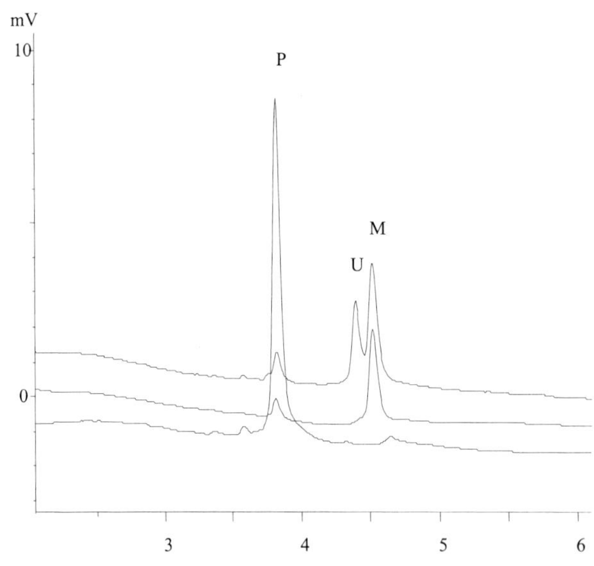

Figure 8.8. The separation of primer extension products that arise from a specific nucleotide sequence that is methylated (M) compared with an unmethylated template. The products of a template in which only one of two alleles is methylated is shown since this is often found in imprinting disorders. Such templates are derived as described in the text as a result of differential bisulfite mediated deamination.

sence of PCR products and nucleotide sequencing is circumvented. The precise difference in methylation at single nucleotide resolution is possible using this method. Since such subtle differences are known to be associated with imprinting diseases, we believe the use of DNA Chromatography in the molecular diagnostics of such diseases will become widespread.

8.4 Nucleic Acid Enzymology

Nucleic acid enzymology (the investigation of enzymes such as DNA polymerases, restriction endonucleases and reverse transcriptases as well as ribozymes such as the hammerhead ribozyme) is dependent upon both analytical and preparative methodologies for the investigation of substrate specificity, post-reaction product characterization and other studies.

DNA Chromatography when coupled with fluorescence detection represents a generic platform for many aspects of nucleic acid enzymology as exemplified below by assays of telomerase, uracil DNA glycosylase and polynucleotide kinase [13]. However, this approach is not confined to these reactions. Indeed, the ability to perform a variety of non-radioactive assays with throughput times of 10 minutes per sample in conjunction with in line data analysis software represents a new direction for DNA Chromatography.

8.4.1 Telomerase Assays

The telomeric repeat amplification protocol (TRAP) is widely employed for the detection of telomerase activity in cell extracts. An alternative fluorescence based modification of the TRAP assay has recently been developed that allows sensitive, high throughput, automated measurement of telomerase activity by DNA Chromatography [13].

Telomerase is a ribonucleoprotein complex that plays a critical role in telomere maintenance and cellular immortality [14]. The vertebrate enzyme catalyzes the addition of hexanucleotide repeats (5′-TTAGGG-3′) to the ends of chromosomes. In the absence of telomerase, human telomeres undergo progressive shortening with each round of cell division, an event that may contribute to cell senescence and mortality. Telomerase is known to be associated with immortalized cancer cells but is absent in most normal tissues and has therefore become a focus for diagnostic investigation. By modifying the PCR-based TRAP assay, a semi-quantitative assay for telomerase activity has been developed that allows high throughput screening of this enzyme with high sensitivity in an automated format [13]. This procedure involves the detection of extension products from a fluorescently, end labeled primer using DNA Chromatography (see Figure 8.9). Telomerase activity is quantified after a TRAP assay by measuring the turnover of the enzyme reaction via an end labeled primer. End labeling the substrate primer ensures the total prod-

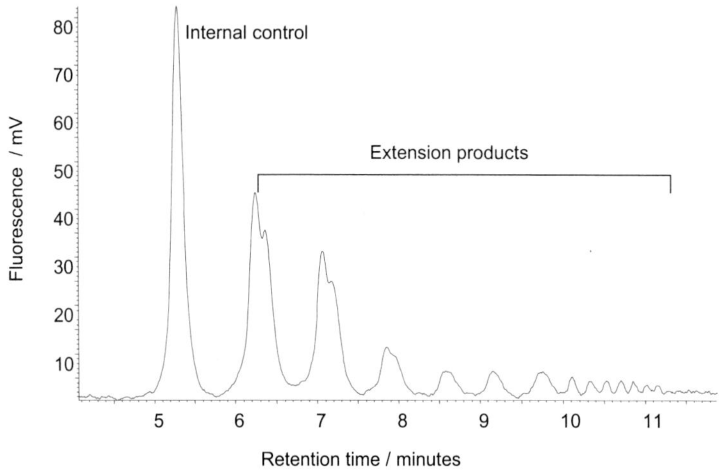

Figure 8.9. The use of DNA Chromatography in size based mode is used analyze the products of the Telomerase reaction leading to the extension of a set of hexanucleotide motifs at the telomeres of chromosomes. As described in the text, the TRAP assay that involves PCR based amplification of the products of telomerase, can be monitored by DNA Chromatography. The extension products therefore have a chromatographic appearance similar to a diminishing sine wave. By comparison with methods in which the products are evaluated collectively, this approach provides both quantitative and qualitative data.

uct signal strength (fluorescence) directly corresponds to the number of substrate primers extended by telomerase and is independent of the length of each individual product. Using the TRAP assay the return primer may bind at varying positions along the extended telomeric sequence, resulting in amplified products that do not directly correspond to the number of telomeric repeats present in the original extended sequence. The assay also incorporates the use of an internal control to allow semi-quantitative analysis of telomerase activity and also reveals the presence of any inhibitors of the PCR reaction, which may produce false negatives.

Using the chromatographic TRAP assay, telomerase activity can be quantified by integrating the fluorescence of the reaction products and comparing it with the fluorescent signal obtained from a known amount of quantification standard (TSR8). The comparison with a standard allows the level of telomerase activity to be described as an absolute level. Using this system of quantification, the telomerase activity is expressed as TPG (total product generated) where one unit is described as 0.001 pmol of primer extended for at least 3 telomeric repeats by telomerase present in the extract. One TPG unit corresponds approximately to that amount of telomerase activity produced by a single immortalized cell.

Figure 8.10 shows a series of chromatograms generated using varying concentrations of the quantification standard TSR8 in the TRAP assay. From these chromatograms peak recognition and quantification software can be used to compute the standard curve from the quantification of the internal control and the TSR8 extension products. One significant advantage of a chromatographic TRAP assay, is the

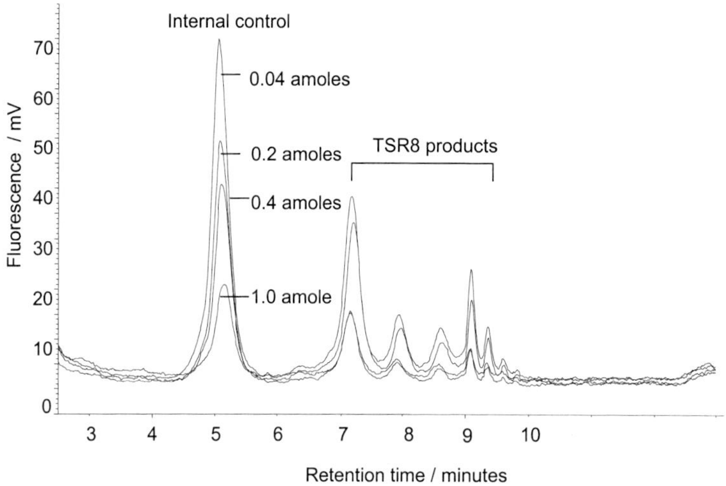

Figure 8.10. The Chromatographic TRAP assay [13] incorporating an internal control duplex that facilitates the accurate quantification of TRAP products as described in the text. The sensitivity of the assay is immediately apparent from the indicated duplex amounts.

ability to evaluate the proportions of individual TRAP products, which may be an indicator of cellular malfunction. Chromatographic TRAP assays may therefore provide additional insight into mechanistic defects in telomerase action, thereby providing not only an alternative diagnostic assay, but also simplifying research into the mechanism of wild type and mutant forms of the enzyme.

8.4.2
Polynucleotide Kinase Assays

Polynucleotide kinase catalyzes the transfer of a phosphate group from ATP to the 5′-OH of DNA, RNA and nucleoside 3′-monophosphates. The enzymatic phosphorylation of 3′ hydroxyl groups of synthetic oligodeoxynucleotides has widespread application in experimental molecular biology. For example, the labeling of oligodeoxynucleotides as probes for identification of target sequences by hybridization is a key technique in molecular cloning [1]. The addition of ^{32}P-labeled phosphate groups to oligodeoxynucleotide sequences using polynucleotide kinase is a common method for creating these labeled probes. Assessment of the proportion of probe that has been labeled and separation of the labeled probe from the unlabelled substrate duplex (or single strand) is normally not undertaken owing to the difficulty of carrying out such procedures.

This subtle ionic difference between phosphorylated and unphosphorylated DNA is readily resolved by DNA Chromatography when the nucleic acid fragment is less than 50 bases in length. This can be demonstrated by separating the product from substrate in a reaction containing an unphosphorylated oligodeoxynucleotide and

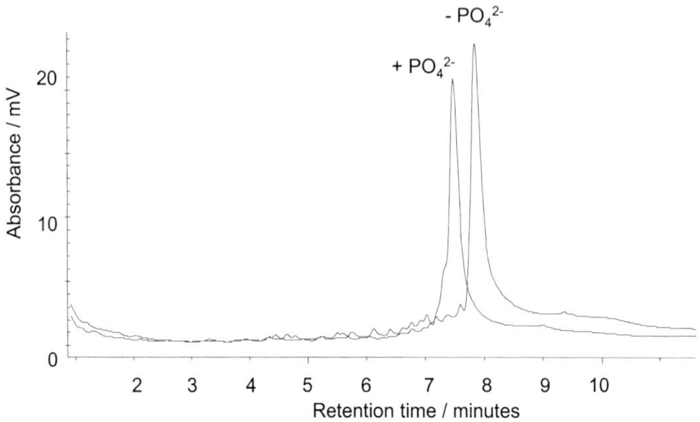

Figure 8.11. The use of DNA Chromatography in the resolution of differentially phosphorylated oligodeoxynucleotides under denaturing conditions. The difference in polarity between the two species is clearly sufficient to resolve the strands, which may be up to 50 (and possibly as much as 100) nucleotides in length. Confirmation of the products was obtained by mass spectrometry [13] which is a powerful complementary technique to DNA Chromatography since the eluted material is ideal for mass spectrometry immediately post column.

polynucleotide kinase. Figure 8.11 shows the effects of phosphorylation of a single-stranded oligodeoxynucleotide upon retention time following DNA Chromatography. Phosphorylation of the oligodeoxynucleotide reduces the retention time of the modified DNA, owing to the addition of the polar phosphate group which diminishes the net hydrophobicity of the DNA. These data are readily corroborated by mass spectrometry.

This particular application of DNA Chromatography demonstrates not only its utility in molecular biology, but also presents for the first time a rapid assay method for exploring the biochemical properties of polynucleotide kinases.

8.4.3 Uracil DNA Glycosylase Assays

Uracil DNA glycosylase (UDGase) is an enzyme that removes uracil from both single-stranded and double-stranded DNA [15]. The enzyme plays an important role in the protection of damaged DNA arising from the deamination of cytosine to uracil. Enzymatic analysis of UDGase is routinely performed by the radio labeling of the DNA strand containing uracil. Following treatment with UDGase the abasic site is heated to 95 °C (often in the presence of base) that results in the cleavage of the labeled strand. This may then be analyzed using denaturing polyacrylamide gel electrophoresis. A novel non-invasive UDGase assay was developed using DNA Chromatography without the need for subsequent cleavage of the DNA strands. Figure 8.12 shows the differential retention time of a DNA strand after treatment with UDGase. It is evident that there is a significant shift in retention time of the modified oligodeoxynucleotide. Removal of the uracil base from the oligodeoxynu-

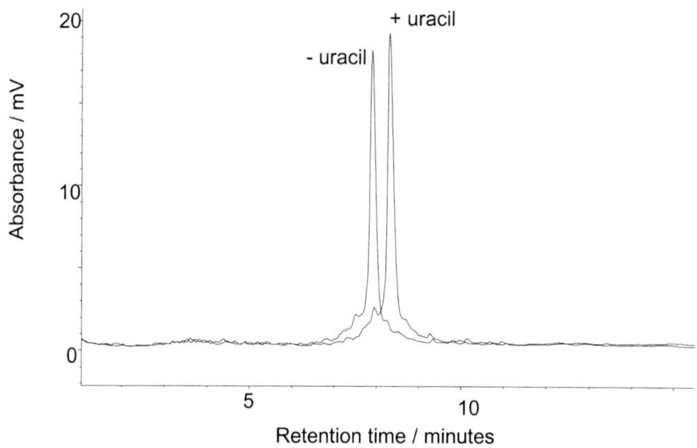

Figure 8.12. The separation of the reactant and product of the reaction catalyzed by uracil glycosylase enzymes. The excision of the base without hydrolysis of the phosphodiester backbone presents a problem in the rapid assay of these enzymes. As shown above, the abasic strand is readily resolved from the substrate by denaturing DNA Chromatography.

cleotide results in the elution at an earlier retention time, as expected through the loss of the uracil base and therefore the subsequent decrease in hydrophobicity of the DNA. This selective reduction in the hydrophobic moment of a uracil containing DNA strand catalyzed by UDGase is discussed with respect to cytosine methylation in Section 8.3.

8.5 Protein Nucleic Acid Interaction Mapping: "Footprinting"

The regulation of expression and the introduction of structural modifications to the genome rely on proteins that bind to specific DNA sequences. The understanding of how such proteins recognize their binding sites in the midst of a plethora of similar sequences in the genome is an important step to understand such processes at the molecular level. Over 30 years ago Galas and Schmitz introduced an elegant method, called "footprinting" to determine the contacts between a specific DNA duplex and a protein [16].

Standard footprinting reactions involve the binding of a protein to radioactively labelled DNA containing the sequence that the protein recognizes. This complex is then digested either enzymatically using DNase I or chemically using hydroxyl radicals [17]. The regions of the DNA molecule covered by the bound protein are protected from digestion while the rest of the DNA backbone undergoes cleavage. The products of the reaction are then separated using electrophoresis; the extent of digestion can then be quantified by further analysis of the gel. A blank region on the autoradiograph is called the footprint and this corresponds to the site of the protein binding to the DNA (see Figure 8.13).

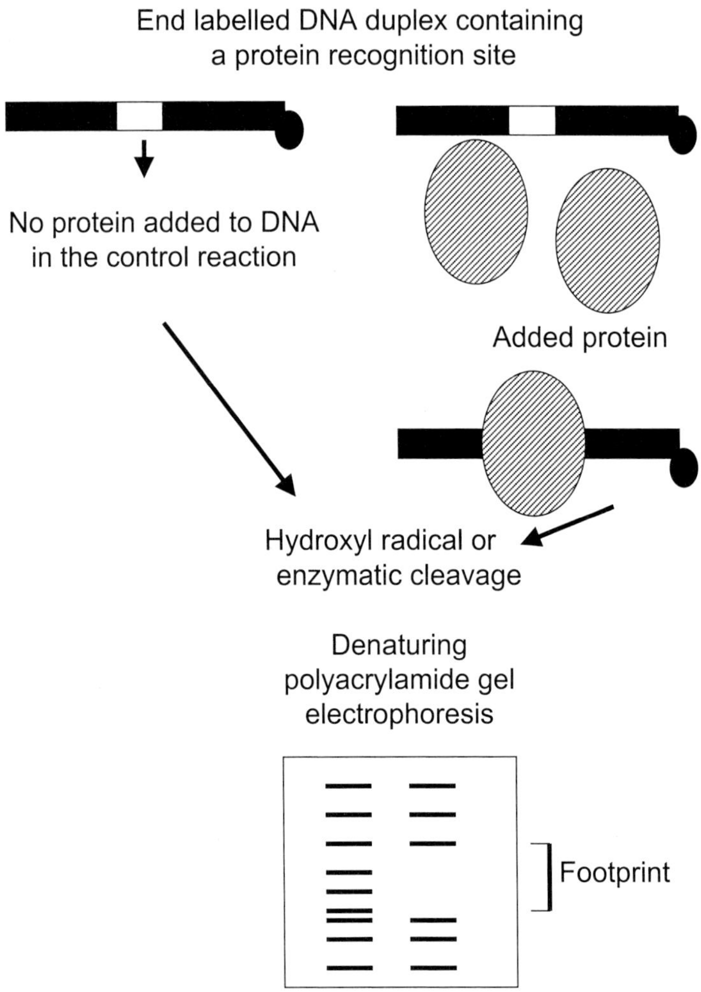

Figure 8.13. Schematic representation of a footprinting experiment to determine the sequence specific interactions between a protein and DNA.

Modifications to the standard footprinting reaction that allow the reaction products to be analyzed by DNA Chromatography include the use of fluorescently labeled DNA. One of the strands of the duplex contains a 5′-fluorescent group introduced during the synthesis of the DNA. This facilitates the analysis of the cleavage products with the aid of a fluorescence detector at the appropriate wavelength(s). Analysis of the footprinting products is rapid with run-times of approximately 30 minutes for each sample. Direct quantification of the cleavage products is performed using peak evaluation software. The analysis of a protein–DNA interaction using hydroxyl radical footprinting is exemplified here by the interaction between the RuvA component of the bacterial resolvasome and a synthetic 4-stranded Holliday junction [18].

Before discussing footprinting data, it is necessary to consider the choice of ion pair reagent and the methodology associated with "phasing" the footprint, that is relating the chromatogram obtained in the presence of the protein to the nucleotide sequence of the DNA duplex. The use of tetrabutylammonium bromide (as the ion pairing reagent) with DNA Chromatography is essential for the size dependent separation of fluorescently labeled DNA. This regime removes the influence of the hydrophobic fluorescent group and any residual sequence specific effects [19]. The sequencing reactions performed involve an express protocol for the generation of a "GA ladder" [20]. This method does not provide complete sequence information; instead it produces cleavage at every G and A residue within the oligodeoxynucleotide strand. However, the Maxam–Gilbert sequencing protocol for complete sequence identification may also be performed, as the chemistries are very similar [20]. It should be noted that the Maxam–Gilbert procedure can be very time consuming and also requires use of high-grade chemicals in order to obtain good results. The express protocol was chosen as a complement to footprinting because it offers a rapid means of obtaining sequence information from a known sequence of DNA. Simultaneous nucleotide sequencing is required to allow complete identification and phasing of all the peaks in the footprinting reaction. The footprinting reaction can generate over 40 chromatographic peaks and it is important to correctly identify each peak (which can often be difficult in the presence of an injection and salt peaks at the start of a chromatogram). Overlaying the GA sequence ladder allows complete characterization of the footprinting reaction once the sequence of the GA ladder has been determined. Some small peaks arise in the sequencing reaction due to small amounts of cleavage at TA residues; however, these anomalous products can be minimized by optimizing the reaction conditions.

Method Section

A typical procedure for express GA cleavage is as follows:

1) DNASep spin columns (experimental spin columns courtesy of Transgenomic, Inc., Omaha, NE) are first incubated with 500 µl of 0.0025 M tBuBr (Tetrabutylammonium bromide)
2) An equal volume of 0.0025M tBuBr was added to the reaction mixtures which are then loaded onto the spin column (typically in a final volume of 100 µl)
3) The column is then washed twice with 0.0025 M tBuBr containing 2 mM EDTA (pH 8.0)
4) The oligodeoxynucleotide fragments are then eluted using 70% acetonitrile.
5) Samples are then lyophilized to remove residual acetonitrile prior to DNA Chromatography.

The results of the purification of a GA sequencing ladder using spin columns described in the methodology section is shown in Figure 8.14. A high concentration of acetonitrile is required to remove the majority of the nucleic acids from the col-

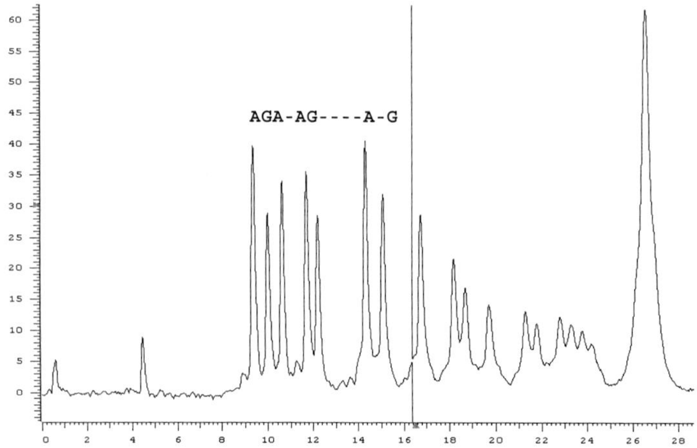

Figure 8.14. The results of a Maxam–Gilbert nucleotide sequencing reaction analyzed by DNA Chromatography under denaturing conditions [13]. The G + A reaction products are detected through strand-specific fluorescent labelling. Part of the nucleotide sequence is written above the peaks.

umn: only a small loss of DNA results from this procedure. Pre-chromatography purification of the reaction products from a chemical cleavage reaction (in the presence and absence of proteins) serves to remove excess reagents (such as iron compounds) and proteins, which can interfere with the resolution of the DNA cleavage products. When six consecutive samples were loaded after purification on the spin columns' no loss in resolution occurred, indicating that the iron salts and protein in the reaction mixture either washed off the column or remain bound after elution.

Purification of complex reaction mixtures containing short oligodeoxynucleotides by DNAsep reverse phase spin columns represents an excellent alternative to similar products from the other commercial suppliers. The above procedure was used to purify a series of nucleic acid fragments from 1–50-mers and can be used to separate radiolabeled DNA fragments from dNTPs (including small oligodeoxynucleotide <15-mers).

The Holliday junction and Holliday junction-RuvA complexes that had been subjected to hydroxyl radical cleavage were analyzed under denaturing conditions at 75 °C using DNA Chromatography (Figure 8.15). The chromatogram of the hydroxyl radical cleavage products from the Holliday junction in the presence of RuvA compares sharply with the cleavage products generated in the absence of the protein (compare Figs. 8.15a and b). Using peak analysis software, the direct quantification of the cleavage products can be achieved by determining the areas under the peaks in each chromatogram. This technique also facilitates the identification of the individual peaks by alignment with fragments from a GA sequencing ladder (see Figure 8.15c). Previous Dnase I footprinting experiments have shown that RuvA binds to the Holliday junction, producing a symmetrical footprint, extending approximately 13 bases on either side of the crossover point [18]. Hydroxyl radical footprinting analysis of the RuvA-Holliday junction complex using this novel fluorescent based analysis determined a more precise footprint, covering approximately 9–10 bases either side of the crossover point. The results from the hydroxyl radical footprinting experiment is in excellent agreement with crystal structures of the RuvA-Holliday junction which show a tetrameric RuvA extending over 8 bases of the crossover point of the Holliday junction.

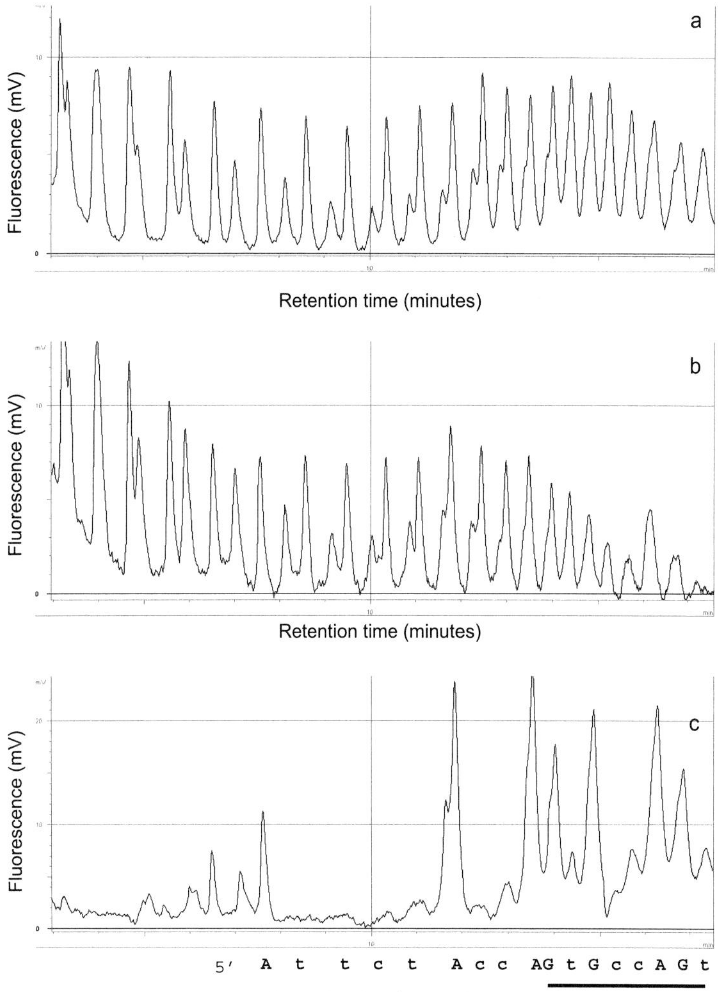

Figure 8.15. The comparative analysis of a DNA four way junction in the absence of the bacterial recombination protein RuvA (a), in the presence of the protein (b) and a corresponding Maxam–Gilbert G + A reaction (c). This experiment shows that hydroxyl radical cleavage is inhibited by the presence of the RuvA protein (solid line underneath chromatogram c, which is an indirect measure of the specific bases that are in contact with the protein.

8.6 Nucleic Acid Tagging

Single-stranded DNA (ssDNA) is an extremely important reagent in molecular biology. The purification and isolation of ssDNA is a key step in numerous analytical molecular biology procedures, including strand-specific hybridization [1], the *in vitro* selection of ssDNA aptamers [21, 22], nucleotide sequencing [1], and molecular weight analysis of DNA using mass spectrometry [23]. Various methods have been used previously to obtain ssDNA from double-stranded (ds) PCR products [24–26]. Importantly most of the techniques require more than one step to isolate the purified ssDNA from PCR products. Here we show that the separation of biotinylated ssDNA from the non-biotinylated ssDNA can be achieved using DNA Chromatography (see Figure 8.16). This allows the isolation of the ssDNA species directly from the dsPCR products in less than 15 min [27]. The PCR is performed using one tagged primer. The tag may be any moiety that confers additional hydrophobic character to one of the duplex strands, thereby facilitating the separation of the ssDNA species under denaturing conditions. Such tags include fluorescent moieties such as 6-FAM or biotin. Other tags that may be used include ones containing positively charged side groups that will also affect the retention time of the ssDNA and therefore will be resolved from the "non-tagged" strand. The rapid isolation of pure ssDNA in high yield is an essential step when performing *in vitro* selection of ssDNA aptamers and the isolation of fluorescently labeled ssDNA is important in the generation of ssDNA probes. The isolation of ssDNA using denaturing DNA Chromatography also facilitates direct analysis by mass spectrometry [23].

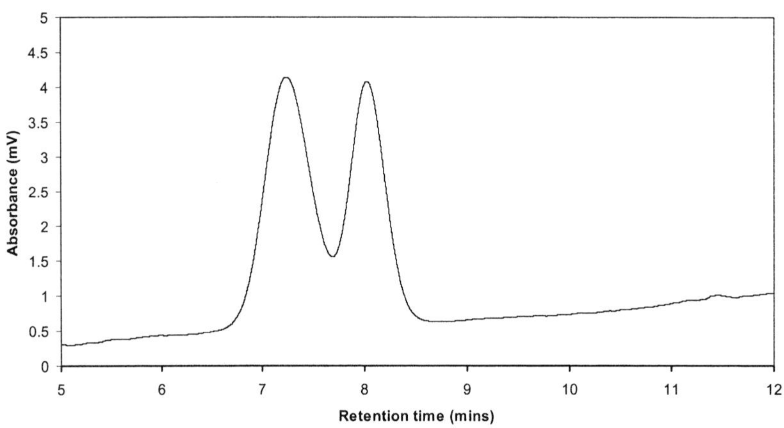

Figure 8.16. Two pools of DNA, one biotinylated and one not are generated in SELEX experiments. DNA Chromatography can be used in order to separate the two pools rapidly and in a preparative mode, thereby streamlining this otherwise cumbersome step in this experimental procedure [26].

8.7
DNA Chromatography with Intercalating Dyes

The sensitivity of nucleic acid detection is a key component of any mutation screening method, and in nucleic acid analyses in general. In high sensitivity work, the use fluorescent dyes have become the preferred means of detecting nucleic acids. Currently, fluorescence based detection generally requires that PCR products are generated by use of fluorescently labelled primers. When single genes are the subject of analysis, the use of fluorescently labelled primers is cost-effective. However when multiple analyses are required, the cost of large numbers of specific fluorescent primers becomes prohibitive unless "universal" tagged primers are utilized.

Fortunately, a simple, inexpensive method of "tagging" DNA by intercalation for detection by DNA Chromatography has been discovered. The method involves simply adding the intercalating reagent to the eluent. Some limited work has also been

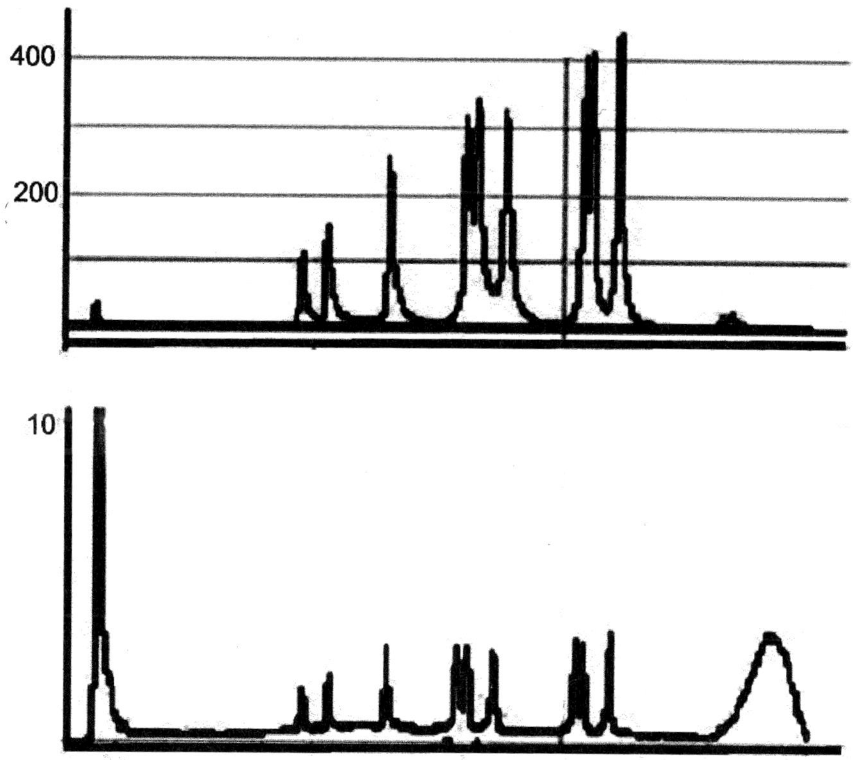

Figure 8.17. The incorporation of SYBR Green intercalating dye into the typical size based DNA Chromatography buffers at a dilution of 1/1000 (see text) is sufficient to provide an increase in the sensitivity of detection of around 50–100 fold without reducing resolution of the DNA fragments. The sensitivity enhancement appears to be greater when the fragments are greater than 100 bp.

done where the reagent has been added to the sample prior to injection. The reagent associates with the fragments as it chromatographs and elutes from the column making the fragment detectable by fluorescence.

The physico-chemical properties of cyanine dye SYBR Green are described in the work at Molecular Probes [28]. Because of the high fluorescence properties experiment work was performed with this dye [29]. In the experiment, SYBR Green at a concentration of 1/1000 of the commercially supplied strength was added to the buffers. The results show (Figure 8.17) that the dye does not interfere with chromatography of DNA and gives a 50–100 fold improvement in sensitivity of detection over conventional UV detection.

References

1. J. Sambrook, E. F. Fritsch and T. Maniatis, in Molecular Cloning: A Laboratory Manual, Second Edition, Cold Spring Harbour Laboratory Press, New York: 1989.
2. M. M. Matin, and D. P. Hornby, A positive selection vector combining tetracycline resistance that eliminates the need for bacterial plating. Anal. Biochem., 278, 46, 2000.
3. A. Pingoud and A. Jeltsch, Structure and function of type II restriction endonucleases, Nucleic Acids Res., 29, 3705, 2001.
4. C. A. Oleykowski, C. R. Bronson Mullins, A. K. Godwin and A. T. Yeung, Mutation detection using a novel plant endonuclease Nucleic Acids. Res., 26, 4597, 1998.
5. D. T. Gjerde and P. D. Taylor, Chromatographic method for mutation detection using mutation site specifically acting enzymes and chemicals. U. S. Pat. 6,027,898, 2000.
6. E. C. Friedberg, G. C. Walker and S. Wolfram, DNA Repair and Mutagenesis, ASM Press: Washington DC, 1994.
7. W. Reik, and J. Walter, Genomic imprinting: parental influence on the genome, Nat. Rev. Genet., 2, 21, 2001.
8. A. P. Bird, CpG-rich islands and the function of DNA methylation, Nature 321, 209, 1986.
9. M. Frommer, L. E. McDonald, D. S. Millar, C. M. Collis, F. Watt, G. W. Grigg, P. L. Molloy and C. L. Paul, A genomic sequencing protocol that yields a positive display of 5-methylcytosine residues in individual DNA strands, Proc. Natl. Acad. Sci. USA, 89, 1827, 1992.
10. J. G. Herman, J. R. Graff, S. Myohanen, B. D. Nelkin and S. B. Baylin, Methylation-specific PCR: A novel assay for methylation status of CpG islands. Proc. Natl. Acad. Sci. USA, 93, 9821, 1996.
11. H. Angelman, 'Puppet children': a report of three cases, Dev. Med. Child Neurol, 7, 681, 1965.
12. B. Hoogendoorn, M. J. Owen, P. J. Oefner, N. Williams, J. Austin and M. C. O'Donovan, Genotyping single nucleotide polymorphisms by primer extension and high performance liquid chromatography, Human Genetics, 104, 89, 1999.
13. M. J. Dickman, M. M. Matin and D. P. Hornby, High throughput analysis of nucleic acid modification reactions using IP RP HPLC. Anal. Biochem. (in press) 2001.
14. G. B. Morin, The human telomere terminal transferase enzyme is a ribonucleoprotein that synthesizes TTAGGG repeats, Cell, 59, 521, 1989.
15. A. M. Delort, A. M. Duplaa, D. Molko, R. Teoule, J. P. Leblance and J. Laval, Excision of uracil residues in DNA – mechanism of action of Escherichia coli and Micrococcus-luteus uracil-DNA glycosylases, Nucleic Acids Res. 13, 319, 1985.
16. D. J. Galas and A. Schmitz, DNase footprinting: a simple method for the detection of protein-DNA binding specificity. Nucleic Acids Res., 5, 3157, 1978.
17. T. D. Tullius, Hydroxyl radical footprinting – high resolution information about DNA

protein contacts and application to lambda-repressor and cro-protein, Proc. Natl. Acad. Sci. USA, 83, 5469, 1990.
18. D. Hargreaves, D.W. Rice, S.E. Sedelnikova, P.J. Artymiuk, R.G. Lloyd and J.B. Rafferty, Crystal structure of E.coli RuvA with bound DNA Holliday junction at 6Å resolution, Nat. Struct. Biol., 6, 441, 1998.
19. P.J. Oefner, C.G. Huber, F. Umlauft, G.N. Berti, E. Stimpfl and G.K. Bonn, High-resolution liquid chromatography of fluorescent dye-labeled nucleic acids, Anal. Biochem., 15, 39, 1994.
20. S. Belikov, and L. Wieslander, Express protocol for generating G+A sequencing ladders, Nucleic Acids Res., 23, 310, 1995.
21. L.C. Bock, L.C. Griffin, J.A. Latham, E.H. Vermass and J.J. Toole, Selection of single stranded DNA molecules that bind and inhibit thrombin, Nature, 355, 564, 1992.
22. A.D. Ellington and J.W. Szostak, Selection in vitro of single stranded DNA molecules that fold into specific ligand binding structures, Nature, 355, 850, 1992.
23. P.A. Limbach, P.F. Crain and J.A. McCloskey, Characterisation of oligonucleotides and nucleic acids by mass spectrometry, Curr. Opin. Biotechnol., 6, 96, 1995.
24. B. Kaltenboeck, J.W. Spatafora, X. Zhang, K.G. Kousoulas, M. Blackwell, and J. Storz, Efficient production of single stranded DNA as long as 2 kb for sequencing of PCR amplified DNA. BioTechniques, 12, 164, 1992.
25. K.P. Williams and D.P Bartel, PCR products with strands of unequal length, Nucleic Acids Res., 23, 4220, 1995.
26. R .G. Higuchi and H. Ochman, Production of single stranded DNA templates by exonuclease digestion following the polymerase chain reaction, Nucleic Acids Res., 17, 5865, 1989.
27. M.J. Dickman and D.P. Hornby, Isolation of single-stranded DNA using denaturing DNA chromatography, Anal. Biochem. 284, 164, 2000.
28. Molecular Probes www.probes.com
29. M.J. Dickman, A. Bahrami, M.M. Matin, J. R . Ashby, M. Conroy,P.E. Brown, G. Fowler, and D.P. Hornby, The incorporation of intercalating dyes for enhancing ion pair reverse phase DHPLC of nucleic acids (submitted) 2001.

9
Looking Forward

Molecular biology is still in the early stages of development. Most causes and controls of cellular functions are yet to be determined. Nevertheless, many genes and their functions have been discovered. The first benefits of molecular biological research have been realized, and more benefits are coming in the form of new drugs. Biotechnology and its derivatives will be one of the major drivers in the world's economy.

The fountain of youth has been a quest of many throughout recorded history. Faced with the choice of looking or feeling younger vs. growing old, the individual will choose youth. Faced with the choice being cured of a disease vs. succumbing to the disease, the individual will choose health. It is certain that appreciable amounts of money will be spent on health and much of this money will be directed toward research tools. If money is spent, technology will evolve.

Given that there will be resources to develop technology what is the future for DNA Chromatography? Much of this depends on the technology drivers. What tools are available to improve the technology? What are the limitations of the technology? Much also depends on the market place and the needs of the researcher. What tools are needed? Market drivers include capability, cost, ease of use, performance, accuracy, precision, sensitivity, and integration with other technologies. Predicting the future is always difficult, but some predictions can be made.

The instrumentation will become smaller. Just as consumer electronic devices have become smaller over the years, analytical instrumentation will also become smaller. The researcher has run out of bench top space and given instruments with equal capability, the researcher will choose the instrument with the smaller footprint. More capability can be packed into the same size package. Miniaturization and integration of components can increase performance (more about this later).

It is usually (but not always) less expensive to manufacture smaller instrumentation. HPLC instrumentation has been shrinking for years and there is no reason why DNA Chromatographic equipment should not do the same. Considerable R&D investment in design and engineering is required to make things smaller, but it is usually well worth it.

Separations will be faster. How do we know separations can be faster? Because in many respects the current instrumentation is actually undoing much of the chromatographic separation process of the column packing material. On a kinetic basis, molecular interaction with the chromatographic surface and mobile phase is a fast process. But extra column effects broaden the peak and work against the resolving power of the column and eluent chemistry. These are normal phenomena that happen with all chromatographic instrumentation. To compensate, conditions are chosen to enhance the differences between materials that are to be separated. This in turn lengthens the separation time. Miniaturization, improved fittings fluid path designs, and new packing methods can all improve the chromatographic process. That is not to say that making chromatography fast is simple. It is only to say that it can be done. How and when are yet to be answered.

The effective cost of analysis will go down. If the PC model is followed, that doesn't mean that the dollar amount spent will go down, but more likely more capability will be the same amount money can be available.

The technology will continue to be integrated with other technologies. The integration may be physical where instrumentation is actually connected with each other. But more likely, papers will continue to be published on collection and downstream processing of materials. The publications show how the technology fits in with the researcher's goals and how these goals can be accomplished. One area where much might be published is RNA analysis.

What tools will be used for nucleic acid analysis? Will chip technology or other technologies dominate? Modern Chromatography is still less than 100 years old. HPLC is 30 years old and modern DNA Chromatography is perhaps only 6 years old. Separation science is fundamental. It seems there will always be the need to prepare and process pure nucleic acids. There will always be the need for a versatile device that can match the scientist's ability to conceive of new research and new experiments. Will DNA Chromatography be part of this? The authors are convinced it will.

Appendix 1
Glossary of Terms

Adapter: A union with different threads on each end; generally used to connect two different types of tubing together.

Adenine (A): A nitrogenous base, one member of the base pair A – T (adenine-thymine).

Alleles: Alternative forms of a genetic locus; a single allele for each locus is inherited separately from each parent (e. g., at a locus for eye color the allele might result in blue or brown eyes).

Amino acid: Any of a class of 20 molecules that are combined to form proteins in living systems. The sequence of amino acids in a protein provides the biological function and is determined by the gene.

Amplification: An increase in the number of copies of a specific DNA fragment; can be *in vivo* or *in vitro*. See cloning and polymerase chain reaction (PCR).

Analytical Column: Typically, the main column used in the DNA Chromatographic system to separate sample components; often referred to as the "heart" of the DNA Chromatograph.

Anneal: The act of binding two single strand of nucleic acid regions of complementary sequences by hydrogen bonding.

Asymmetry factor: A factor used to describe the shape of a chromatographic peak. While theory assumes a symmetrically shaped peak, in practice the peak is often asymmetric. The equation used to describe this factor is determined at 10 percent of the peak height. At this level, the factor is defined as being the ratio of the distance between the peak's maximum and the back of the peak's curve to the distance between the peak's maximum and the front of the peak's curve.

Autoradiography: A technique that uses X-ray film to visualize radioactively labeled molecules or nucleic acid fragments after they are separated by gel electrophoresis.

Autosampler: Automatically takes sample from the PCR plate, places the sample in the injector valve and injects the sample into the mobile phase flowing to the column.

Back pressure: A phrase used to describe the pressure required to force the eluent at a determined flow rate along a system's flow path, typically expressed in psi, bar, atm, or MPa.

Back Pressure Regulator: A device typically used after the detector to maintain a positive pressure on the flow cell, thus minimizing solvent out gassing problems in the detector. It can also be used as a PUMP PRELOAD, to help the pump's check valves operate more efficiently.

Band: A pattern of dark regions on a gel by staining that indicate the presence of a nucleic acid. Or a visual deflection along a chromatogram's baseline, representing a sample component, also known as a peak.

Bandspreading: Refers to the dilution of the chromatographic band as it moves from the injection valve through the column and to the detector.

Band width: The width of the chromatographic band measured at the baseline. The band width indicates how efficient the column is at separation. It also indicates how many peaks can be separated in one run i. e. the peak capacity.

Bar: A unit of pressure equal to one atmosphere; also equivalent to about 15 pounds per square inch (psi) or 0.1 Megapascal.

Baseline: The chromatographic baseline is the line drawn by the recording device representing the signal from the detector when only mobile phase or eluent is being pumped through the system. It also represents the point from which peak area or peak height is measured.

Baseline noise: The variation along the chromatographic baseline due to electrical noise or temperature fluctuations, but can also be due to out gassing (bubble formation) in the detector flow cell (seen as baseline spikes), as well as poorly mixed gradient.

Base pair (bp): Two nitrogenous bases (adenine and thymine or guanine and cytosine) held together by hydrogen bonds. Two strands of DNA are held together in the shape of a double helix by the bonds between base pairs.

Base pairing: The pairing of specific nitrogenous bases between complementary strands of DNA. For example, adenine is always paired with thymine and guanine with cytosine.

Base sequence: The order of nucleotide bases in a DNA molecule.

Base sequence analysis: A method, sometimes automated, for determining the base sequence.

Binary solvent gradient: A gradient composed of two solvents being mixed together. Most gradients are of this type.

Biocompatible: Refers to that special property that allows some materials (metals plastics, etc.) to come into contact with biological materials without changing or trapping the biological material. Biocompatibility does not necessarily mean the material is DNA compatible. See DNA compatible.

Biotechnology: A set of biological techniques developed through basic research and now applied to research and product development.

bp: See base pair.

Cancer: A disease process initiated by transformation of a cell to behavior lacking in normal growth controls often involving invasive behavior.

Capacity factor (k or k'): A way of expressing the retention of the sample components. It is determined by subtracting the column void time from the retention time of the peak of interest, and dividing that by the column void time.

Capillary gel electrophoresis (CGE): A type of gel electrophoresis the uses voltage to selectively separate charged nucleic acids in a capillary filled with a buffer and a linear polymeric solution. It is an automated technique that produces sharp bands with minimal tailing

Capillary electrochromatography (CEC): A type of chromatographic analysis which combines electrophoresis and capillary HPLC. Rather than a pump, the eluent flow through the system is produced using electro osmotic flow (EOF). Molecules flow through the system with this EOF "pump".

Cartridge column: A guard or analytical column that comes pre-packed, and is placed in a reusable holder.

cDNA: See complementary DNA.

Channeling: A phenomenon occurring when voids or bed fractures occur in the column bed, often due to drying of the column bed or dropping the column.

Check Valve: A device inserted into a moving liquid stream that allows flow of the stream in only one direction; often built into the HPLC pump. Each pump has at least one each of inlet and outlet check valves.

Chromatogram: A plot of a detector's signal output vs. time. It is identified by a series of peaks or bands on a baseline.

Chromatography: Literally "color writing," a term describing a separation technique that occurs based upon the difference of interaction of sample components with two phases: These are a stationary phase or a column and liquid mobile phase or eluent.

Chromatographic conditions: Those parameters that describe how an analysis was achieved for the purpose of potential future duplication for verification purposes.

Chromosomes: The self-replicating genetic structures of cells containing the cellular DNA that bears in its nucleotide sequences in a linear array of genes. In prokaryotes, the DNA is circular, and the entire genome is carried on one structure (or chromosome). Eukaryotic genomes are carried on DNA structures that may be circular or linear.

Clone bank: See genomic library.

Clone (noun): One of a collection of cells or vectors containing identical genetic material.

Clones: A group of cells derived from a single ancestor.

Clone (verb): The act of duplicating genetic material within a vector.

Cloning: The process of asexually producing a group of cells (clones), all genetically identical, from a single ancestor. In recombinant DNA technology, the use of DNA manipulation procedures to produce multiple copies of a single gene or segment of DNA is referred to as cloning DNA.

Cloning vector: DNA molecule originating from a virus, a plasmid, or the cell of a higher organism into which another DNA fragment of appropriate size can be integrated without loss of the vectors capacity for self- replication; vectors introduce foreign DNA into host cells, where it can be reproduced in large quantities. Examples are plasmids, cosmids, and yeast artificial chromosomes; vectors are often recombinant molecules containing DNA sequences from several sources.

Code: The system by which a nucleic acid sequence made up of 4 different bases will specify a particular protein to be produced. It takes a combination of 3 nucleotides to specify 1 amino acid, and there are 64 codons of which three specify the termination of protein synthesis (the process usually referred to as translation). [Protozoa have a variation on the Universal genetic code in that a termination codon is utilised to specify incorporation of the amino acid Trp] The Code is also referred to as the genetic code A protein typically comprises between 100 and 1000 amino acids, although there are some shorter and some longer polypeptide chains found in Nature. The genetic code is not to be confused with the genome, which is by definition the complete set of chromosomes required to specify a given organism.

Codon: A group of three consecutive nucleotides within messenger RNA (mRNA) that encodes a message to initiate translation to incorporate a specific amino acid into the growing polypeptide chain, or to stop translation.

Column: Typically a tube of sorts that contains the stationary phase or column packing of a chromatographic system.

Column dead time: The time it takes for solvent molecules or unretained sample bands to pass through the column unhindered through the column. Uracil is often used to measure the column dead time in a reverse phase column. See dead time.

Column packing: The particulate material packed inside the column; also called the stationary phase. This usually consists of non porous, polymer-based particles chemically bonded with a C18 chemical functional group.

Column performance: A column's efficiency, often expressed as a height equivalent theoretical plate value, used to describe how well a column will resolve sample fragments.

Concentricity: In HPLC, the degree to which the through-hole of a part has a common centerline with the part. This is critical when aligning small-bore tubing with a union or column through-hole. If the through-hole is not concentric, the flow path may be restricted.

Controls: See internal standards and external standards.

Complementary DNA (cDNA): DNA that is synthesized from a messenger RNA template; the single-stranded form is often used as a probe in physical mapping.

Complementary sequences: Nucleic acid base sequences that can form a double-stranded structure by matching base pairs; the complementary sequence to G – T – A – C is C – A – T – G.

Consensus sequence: A term that refers to sequences common to different genes within an organism, or to the same gene among different organisms, that encodes a specific function. This term may be applied to either nucleic acids or proteins, since the protein sequence is completely dependent upon the nucleic acid sequence.

Conserved sequence: A base sequence in a DNA molecule (or an amino acid sequence in a protein) that has remained essentially unchanged throughout evolution.

Crossover: The breaking during meiosis of one maternal and one paternal chromosome, the exchange of corresponding sections of DNA, and the rejoining of the chromosomes. This process can result in an exchange of alleles between chromosomes. Compare recombination.

Cytosine (C): A nitrogenous base, one member of the base pair G–C (guanine and cytosine).

Dead time: Also known as hold up time, the amount of time it takes to sweep the dead volume of the mobile phase through the system for a given flow rate. This is different from the gradient delay time. See dead volume.

Dead volume: The volume of the flow path from the injection valve to the detector cell. These small spaces within the LC system allow remixing of the separated sample bands and should be minimized in a DNA Chromatographic system. This is different from the gradient delay volume See dead time.

Degassing: The process of removing dissolved gases from the mobile phase or eluent to prevent corrosion and to provide stable detection results. In DNA Chromatography, degassing is achieved through inline vacuum degassing.

Degeneracy: In molecular biology this term refers to the fact that multiple different codons may encode the same amino acid. However, a given codon does not encode more than one amino acid within the nucleus of an organism.

Deletion: The absence of bases that are present in the wild-type DNA sequence. See insertion.

Deoxyribonucleotide: See nucleotide.

Detector: One of the basic components of an DNA Chromatographic system; responsible for providing the presence and amount of separated sample components as they pass through the flow cell.

Detector background signal: Sometimes called the detector background noise, the signal given by a detector when no sample is present. The background signal is determined by measuring the peak to peak distance of the baseline with detection set to a high sensitivity. See detection limit.

Detection limit: The amount of sample (in mass or concentration) giving a detector signal that is 3 times the background signal. This should not be confused with detector sensitivity. See detector background signal.

Diploid: A full set of genetic material, consisting of paired chromosomes one chromosome from each parental set. Most animal cells except the gametes have a diploid set of chromosomes. The diploid human genome has 23 pairs of chromosomes. Compare haploid.

DNA (deoxyribonucleic acid): The molecule that encodes genetic information. DNA is a double strand molecule held together by weak bonds between base pairs of nucleotides. The four nucleotides in DNA contain the bases: adenine (A), guanine (G), cytosine (C), and thymine (T). In nature, base pairs form between A and T and between G and C; thus the base sequence of each single strand can be deduced from that of its partner.

DNA compatible: Refers to a special property of materials allowing them to come into contact with nucleic acids without changing or trapping the material. Biocompatibility usually refers to proteins and does not necessarily mean the material is DNA compatible.

DNA denaturation or melting temperature: The temperature at which double helix of a fragment of DNA will break their hydrogen bonds and become 50% separated or opened. Depending on the sequence, DNA contains different domains that will melt at different temperatures (temperature where the domain is 50% separated or opened). GC rich domains melt at higher temperatures than AT rich domains.

DNA ligase: Enzymatic activity responsible for creating phosphodiester bonds between the 5' phosphate end of one strand of DNA and the 3' hydroxyl end of another strand or of the same strand. Requires the presence of a 5' phosphate on one strand, and a 3' hydroxyl group on the second strand.

DNA polymerase: Enzymatic activity responsible for catalyzing the polymerization of DNA. Is dependent upon an annealed primer from which to initiate polymerization, and a DNA template from which to copy.

DNA probes: See probe.

DNA replication: The use of existing DNA as a template for the synthesis of new DNA strands. In humans and other eukaryotes, replication occurs in the cell nucleus.

DNA sequence: The relative order of base pairs, whether in a fragment of DNA, a gene, a chromosome, or an entire genome. See base sequence analysis.

Domain: A discrete portion of a protein with its own function. The combination of domains in a single protein determines the overall function.

Double helix: The helical shape assumed by DNA in which the two complementary strands hydrogen bond together in opposite orientations (i. e. have opposite polarities or directional sense).

Drain Valve: Otherwise known as a purge valve, it is used to direct the pump fluid stream to waste for flushing or solvent change over.

Dwell volume: See Gradient delay volume.

E. coli: Common bacterium that has been studied extensively by geneticists because of its small genome size, normal lack of pathogenicity, and ease of growing in the laboratory.

Effluent: The liquid mobile phase or eluent that flows from the column or flows from the instrument.

Electrophoresis: A method of separating large molecules (such as DNA fragments or proteins) from a mixture of similar molecules. An electric current is passed through a medium containing the mixture, and each kind of molecule travels through the medium at a different rate, depending on its charge and size. Agarose and polyacrylamide gels are the media commonly used for electrophoresis of proteins and nucleic acids because they have minimum diffusion and separate by size (in the case of nucleic acids).

Endonuclease: An enzyme that cleaves its nucleic acid substrate at internal sites in the nucleotide sequence.

Enzyme: A protein that acts as a catalyst, speeding the rate at which a biochemical reaction proceeds but not altering the direction or nature of the reaction *in vitro.*

EST: Expressed sequence tag. See sequence tagged site.

Eukaryote: Cell or organism with membrane bound, structurally discrete nucleus and other well developed subcellular compartments known as organelles. Eukaryotes include all organisms except viruses, bacteria, and blue-green algae. Compare prokaryote.

Eluate: The combination of the eluent and the sample exiting the column.

Eluent: Another term for the mobile phase used to carry out a separation.

Elution: A term used to describe the passing of eluent through the column for the purpose of moving sample fragments through the instrument.

End fitting: The fitting at the ends of the column that allow the tubing and column bed to interface. Additionally, the end fitting holds the frits in place on both ends of the column, thus retaining the packing material within the column tube.

Evolutionarily conserved: See conserved sequence.

Exogenous DNA: DNA originating outside an organism.

Exons: The protein coding DNA sequences of a gene. Compare introns.

Exonuclease: An enzyme that cleaves nucleotides sequentially from 3' or 5'ends of a linear nucleic acid substrate.

Expressed gene: See gene expression.

Extra-column effects: Those portions of the flow path outside the column itself that contribute to band broadening. Common culprit areas include the injection valve, the detector, the tubing and fittings used along the flow path, frits, etc.

External Standard: A special sample which contains known quantities of the fragments of interest. An external standard is used to help identify sample fragments by comparing the time in which they eluted, to times obtained through the injection of the external standard under the same conditions. An external standard may of known concentration may be used to measure the concentration of the unknown fragment by comparing the relative peak areas or peak heights.

False negative result: A result from an analytical test that is less than the actual amount present. In many cases, a false negative result means something is not measured at all when in fact it is present.

False positive result: A result from an analytical test that is greater than the actual amount present. In many cases, a false positive result means something is measured when in fact it is not present.

Ferrule: A tapered conical ring used to make the seal between a piece of tubing and a receiving port. Ferrules almost invariably must be used in conjunction with an HPLC tubing fitting or is part of the fitting itself.

FISH (fluorescence *in situ* hybridization): A physical mapping approach that uses fluorescent tags to detect hybridization of probes with metaphase chromosomes and with the less condensed somatic interphase chromatin.

Fittings: Refers to the HPLC connectors that join tubing, columns and various LC modules together. The HPLC fitting is made up of a body that has threads to hold the fitting in place and a ferrule to seal the tubing. See ferrule.

Flow rate: Describes amount of eluent flowing through the column and system most often expressed in mL/min (milliliters per minute) for typical DNA Chromatography applications.

Flow cytometry: Analysis of biological material by detection of the light-absorbing or fluorescent properties of cells or subcellular fractions (i. e., chromosomes) passing in a narrow stream through a laser beam. An absorbance or fluorescence profile of the sample is produced. Automated sorting devices, used to fractionate samples, sort successive droplets of the analyzed stream into different fractions depending on the fluorescence emitted by each droplet.

Flow karyotyping: Use of flow cytometry to analyze and/or separate chromosomes on the basis of their DNA content.

Fluorescence detector: A type of detection instrument often used in DNA Chromatography, in which sample fragments containing a tag are exposed to a specific wavelength of light. This exposure causes the sample fragment to "fluoresce," or emit light of a different wavelength that is detected by the fluorescence detector.

Frame-shift: A change from one reading frame to another.

Frit: A porous metal or polymeric filtration component that is used to perform the following: a) prevent particulates from entering the DNA Chromatographic system from the eluent or sample, b) contain the column packing inside the column. Frits used in DNA Chromatographic components normally are 0.5 micron porosity.

Fronting: A non Gaussian curve in which the peak "leads in," and is shown by slight tapering on the front part of the peak's trace. Fronting is the opposite of tailing. See tailing.

Gamete: Mature male or female reproductive cell (sperm or ovum) with a haploid set of chromosomes (23 for humans).

Gaussian curve or peak: Refers to the "perfect" peak characterized by its symmetrical bell shape. A true Gaussian curve is rarely encountered in day-to-day chromatography; however, peaks that deviate greatly from the ideal are an indication either of a problem with the column or eluent or with the flow path.

gDNA: Genomic DNA.

Gene: The fundamental physical and functional unit of heredity. The sequence of nucleotides, coded in triplets (codons) along the mRNA, determines the sequence of amino acids in protein synthesis. The DNA sequence of a gene can be used to predict the mRNA sequence, and can in turn be used to predict the amino acid sequence.

Gene expression: The process by which a gene's coded information is converted into the structures present and operating in the cell. Expressed genes include those that are transcribed into mRNA and then translated into protein and those that are transcribed into RNA (e. g., transfer and ribosomal RNAs).

Gene families: Groups of closely related genes that make similar products.

Gene library: See genomic library.

Gene mapping: Determination of the relative positions of genes on a DNA molecule (chromosome or plasmid) and of the distance, in linkage units or physical units between them.

Gene product: The biochemical material, either RNA or protein, resulting from expression of a gene. The amount of gene product is used to measure how active a gene is; abnormal amounts can be correlated with disease causing alleles.

Genetic code: See code.

Genetic engineering technologies: See recombinant DNA technologies.

Genetic map: See linkage map.

Genetic material: See genome.

Genetics: The study of the patterns of inheritance of specific traits.

Genome: All the genetic material in the chromosomes of a particular organism; its size is generally given as its total number of base pairs.

Genome projects: Research and technology development efforts aimed at mapping and sequencing some or all of the genome of human beings and other organisms.

Genomic library: A collection of clones made from a set of randomly generated overlapping DNA fragments representing the entire genome of an organism. Compare library, arrayed library.

Germline: Refers to germline cells or germline mutations. Germline cells are responsible for passing along the organism's genetic material to the offspring. Germline mutations are hereditary. See somatic.

Gradient delay time: The time needed for the eluent gradient to travel from its start (when the gradient is first mixed) through the mixer and injector to the top of the column. This is different than dead time or holdup time. See gradient delay volume.

Gradient delay volume: The volume of the eluent starting when the gradient is first mixed and continuing through the mixer and injector to the top of the column. This is different than dead volume or holdup volume. See gradient delay time.

Gradient elution: An HPLC procedure where the composition of the mobile phase increases in strength during the separation. This is the opposite of isocratic elution where the mobile phase composition remains constant throughout the entire analysis. A typical example is where the mobile phase starts out as water with a low concentration of acetonitrile and more acetonitrile is added continuously with the separation.

Guanine (G): A nitrogenous base, one member of the base pair G – C (guanine and cytosine).

Guard column or cartridge: A short column or cartridge placed between the sample injector and the inlet of the main column intended to absorb or pick up impurities or particulates in the sample or eluent that might damage the main column and lead to high eluent back pressure.

Haploid: A single set of chromosomes (half the full set of genetic material), present in the egg and sperm cells of animals and in the egg and pollen cells of plants. Humans have 23 chromosomes in their reproductive cells. Compare diploid.

Heteroduplex: A double-stranded fragment of DNA where one or more of the nucleotides are not complementarily matched and are not hydrogen bonded. Two heteroduplex and two homoduplex species each are formed by heating and reannealing of mutated and nonmutated double-stranded fragments. See homoduplex.

Heterozygosity: The presence of different alleles at one or more loci on homologous chromosomes.

Heterozygous: An individual containing dissimilar alleles for a given gene or locus.

High pressure mixing: A pumping procedure in which two or more different solvents are mixed on the high pressure side of the pumps to form a final mobile phase. Normally a pump is needed for each solvent used.

Hold up time: See dead time.

Homozygous: An individual containing identical alleles for a given gene or locus.

Homoduplex: A double-stranded fragment of DNA where the nucleotides are complementarily matched. Two heteroduplex and two homoduplex species each are formed by heating and reannealing of mutated and nonmutated double-stranded fragments. See heteroduplex.

Homologies: Similarities in DNA or protein sequences between individuals of the same species or among different species.

Homologous chromosomes: A pair of chromosomes containing the same linear gene sequences, each derived from one parent.

Human gene therapy: Insertion of normal DNA directly into cells to correct a genetic defect.

Human Genome Initiative: Collective name for several projects started in 1986 by DOE to (1) create an ordered set of DNA segments from known chromosomal locations, (2) develop new computational methods for analyzing genetic map and DNA sequence data, and (3) develop new techniques and instruments for detecting and analyzing DNA. This DOE initiative is now known as the Human Genome Program. The national effort, led by DOE and NIH, is known as the Human Genome Project.

Hybridization: The process of aligning and joining two complementary strands of DNA or one each of DNA and RNA to form a double-stranded molecule. Hybridization is widely used in heteroduplex and homoduplex formation, identification of portions of DNA on a Southern (or Northern) blot using labeled probes, array or chip technology and other processes.

Hydrophilic: Literally water-loving. Refers to compounds, solvents or column surfaces that either dissolve easily in water, or prefer or attract water compared to non-polar organic solvents. See hydrophobic.

Hydrophobic: Literally water-hating. Refers to compounds, solvents or column surfaces that either dissolve easily in non-polar organic solvents such as acetonitrile, or attract or prefer such solvents compared to water. See hydrophilic.

Inflection point: The point in a plot of detector response vs. time at which a line or turns or bends from a direct line to another direction. The point at which a curve turns or changes shape.

Informatics: The study of the application of computer and statistical techniques to the management of information. In genome projects, informatics includes the development of methods to search databases quickly, to analyze DNA sequence information, and to predict protein sequence and structure from DNA sequence data.

Injection solvent: The solvent that the sample is dissolved into before it is injected. This includes solvent from the autosampler and from the sample matrix.

Injector: One of the basic components of DNA Chromatographic system; allows a predetermined amount of sample to be introduced into the eluent flow path to the column. The injector may be a manually-actuated valve, but for most systems, it is completely automated and unattended.

Inlet check valve: The check valve(s) on a pump that allow(s) the mobile phase to flow from the reservoirs into the pump, but not in the reverse direction. See outlet check valve and check valve.

Inlet filters: Filtration devices attached to the inlet lines of the eluent in the reservoirs that remove particulate matter from the eluent before the solvent reaches the pump.

Inline filter: A small volume device containing a frit which prevents particulate matter from the system, eluent or sample from contaminating the system in any way. Often used between the pump and the injector to keep particles shredding off the pump seal surfaces from contaminating the injector valve or used directly in front of the column.

Insertion: The presence of additional bases within a sequence that are not present in wild-type sequence. See deletion.

In situ hybridization: Use of a DNA or RNA probe to detect the presence of the complementary DNA sequence in cloned bacterial or cultured eukaryotic cells.

Interference: An unwanted band that overlaps one of the desired bands in a chromatogram. Interferences reduce the reliability of the results.

Internal standards: Used primarily for the purpose of calibration. Internal standards consist of a known concentration of a sample, different from the actual analyzed sample, which help minimize analytical errors such as detection or volumetric errors that may occur during sample handling or preparation. An internal standard will also minimize sample preparation errors such as PCR or isolation because the sample and standard experience the same environment in the preparation process. See external standards. See controls.

Introns: The DNA base sequences interrupting the protein- coding sequences of a gene; these sequences are transcribed into RNA but are cut out of the message before it is translated into protein. Compare exons.

In vitro: Outside a living organism meaning in the glass (test tube).

Ion exchange chromatography (IEC): A chromatographic method in which eluent ions displace sample ions on anionic or cationic sites within the column. In DNA Chromatography, the column contains positively charged anion exchange sites.

Ion pair chromatography: A chromatographic method in which the sample ions are "paired" with the opposite charged agents, thus forming a neutral ion pair and allowing the sample to be retained on a nonpolar, reverse phase column.

Isocratic elution: A chromatographic separation where the mobile phase composition stays the same during the entire analysis. Isocratic separations are the opposite of gradient elution separations.

Karyotype: A photomicrograph of an individuals chromosomes arranged in a standard format showing the number, size, and shape of each chromosome type; used in low resolution physical mapping to correlate gross chromosomal abnormalities with the characteristics of specific diseases.

kb: See kilobase.

Kel-F® (PCTFE): A chemically resistant fluoropolymer used to make fittings and accessories for fluid transfer applications. It is molecularly related to TEFLON® (PTFE).

Kilobase (kb): Unit of length for DNA fragments equal to 1000 nucleotides.

Library: An unordered collection of clones (i. e., cloned DNA from a particular organism), whose relationship to each other can be established by physical mapping. Compare genomic library, arrayed library.

Linear velocity: Related to the speed at which eluent and thus the sample components are moving within a column. The linear velocity is calculated from the flow rate and the volume per unit length of the column.

Linkage: The proximity of two or more markers (e. g., gene, sequence or mutation) on a chromosome; the closer together that the markers are, the lower the probability that they will be separated during DNA repair or replication processes and hence the greater the probability that they will be inherited together.

Linkage map: A map of the relative positions of genetic loci on a chromosome, determined on the basis of how often the loci are inherited together.

Loop: A piece of tubing connected to a sample injector that holds the volume of sample to be introduced into the flow path.

Low pressure mixing: A pumping procedure that mixes two or more solvents before the pump to form the mobile phase and then delivers the mixture to the system. Most often just a single pump plus a controller is used.

Localize: Determination of the original position (locus) of a gene or other markers on a chromosome.

Locus (plural loci): The position on a chromosome of a gene or other chromosome marker or, the DNA at that position. The use of locus is sometimes restricted to mean regions of DNA that are expressed. See gene expression.

Macrorestriction map: Map depicting the order of and distance between sites at which restriction enzymes cleave chromosomes.

Mass detection limit: The amount of sample mass that must be present to give a detection signal that is 3 times the detector background signal. Most analysis should be well above the limit i. e. 10 times the detector background signal.

Mass sensitivity: The amount of detector signal for a given amount of sample in mass units. See sensitivity.

Mass spectrometry detector: A chromatographic detector in which sample components are ionized and resolved on the basis of their mass to charge (m/z) ratio due to their behavior in electric and magnetic fields.

Mass transfer: Refers to the movement of the solute into and out of the stationary phase and mobile phase; a large mass transfer value indicates a better performing column. A large stationary phase particle size, a low eluent temperature and high eluent viscosity can decrease mass transfer.

MegaPascal (MPa): A unit of pressure; one MPa equals about 10 bar (atmospheres) or 150 pounds per square inch (psi).

Microbore: Describes columns used in applications where the inside column diameter is less than or equal to 2 mm.

Mobile Phase: Also called the eluent, one of the two general "phases" described in an HPLC system. The mobile phase refers to the solvent that is moving (or is "mobile") through the system. See also stationary phase.

Mobile phase strength: Determines how fast the sample moves through the column; a strong mobile phase results in sample bands coming out fast, and a weak mobile phase gives longer retention times for each band.

Molecular biology: The biochemical study of the genetic basis of cells.

Mapping: See gene mapping, linkage map, and physical map.

Marker: An identifiable physical location on a chromosome (e. g., restriction enzyme cutting site, gene) whose inheritance can be monitored. Markers can be expressed regions of DNA (genes) or a short segment of DNA with no known coding function but whose pattern of inheritance can be determined.

Mb: See megabase.

Megabase (Mb): Unit of length for DNA fragments equal to 1 million nucleotides and roughly equal to 1 cm.

Messenger RNA (mRNA): RNA that serves as a template for protein synthesis. See genetic code.

mRNA: See messenger RNA.

Multifactorial or multigenic disorders: See polygenic disorders.

Multiplexing: A sequencing approach that uses several pooled samples simultaneously, greatly increasing sequencing speed.

Mutagen: An agent capable of causing mutations. Common examples are ultraviolet light, such as in sunlight, and anthracene, a material formed during the cooking of fatty meats on a barbecue grill.

Mutation: A permanent, heritable change of the genetic material, either in a single gene or in the numbers or structures of the chromosomes. Compare polymorphism.

Nitrogenous base: A nitrogen- containing molecule having the chemical properties of a base.

Noise: In HPLC, this usually refers to an irregular baseline. All baselines become "noisy" as the sensitivity of the detector is increased.

Non polar: Compounds, solvents or surfaces that dissolve readily in solvents, such as acetonitrile, or prefer such solvents in place of water. Non polar substances do not dissolve in water.

Northern blot: The transfer of size-separated RNA fragments to a synthetic membrane for further studies – like Southern blot but for RNA identification.

Nucleic acid: A large molecule composed of nucleotide subunits.

Nucleoside: A term referring to the combination of adenine, cytosine, guanine, or thymine with a ribose or 2-deoxyribose sugar moiety. A nucleoside is not phosphorylated.

Nucleotide: A monomeric unit of DNA or RNA consisting of a nitrogenous base, adenine, guanine, thymine, or cytosine (dATP, dGTP, dTTP, or dCTP) in DNA; adenine, guanine, uracil, or cytosine (ATP, GTP, UTP, or CTP) in RNA, a phosphate molecule, and a sugar molecule (2-deoxyribose in DNA and ribose in RNA). A nucleotide contains the base attached to a phosphorylated form of 2-deoxyribose or ribose. Thousands of nucleotides are linked to form a DNA or RNA molecule. See DNA, base pair, RNA.

Oligonucleotide: Single stranded nucleic acid, usually DNA, but may also be RNA. In this book will mostly refer to single stranded DNA. The molecule can be of any length, but usually refers to short strands less than 30 mer. The name may sometimes be shorted to oligos. See oligodeoxribonucleotide and oligoribonucleotide

Oligodeoxribonucleotide: Single stranded DNA.

Oligoribonucleotide: Single stranded RNA.

Oncogene: A gene, one or more forms of which is associated with cancer. Many oncogenes are involved, directly or indirectly, in controlling the rate of cell growth.

Overlapping clones: See genomic library.

Outlet check valve: The check valve(s) on a pump that allow(s) the mobile phase to flow from the pump to the autosampler injector, column and detector, but not in the reverse direction. See inlet check valve and check valve.

Packing material: Refers to the stationary phase used in the column.

Partially resolved peaks: Where multiple peaks run together and do not begin and end on the baseline. In order for a peak to be considered completely resolved, it must begin and end on the baseline without interference from other peaks.

Particulate: Small, solid particles in the sample or mobile phase which can plug or damage the column or other parts of the HPLC system.

PCR: See polymerase chain reaction.

PCR induced mutations: Mutations cause by errors as part of the amplification process. Use of a proofreading polymerase for amplifications can reduce the number of mutations.

PCR noise: See PCR induced mutations.

Peak: The visual detector response on the chromatogram based due to the presence of a sample fragment. Also referred to as a band.

PEEK™ (polyetheretherketone): A chemically resistant polymer used to make fittings and tubing.

Phenotype: The physical expression of a gene.

Physical map: A map of the locations of identifiable landmarks on DNA (e. g., restriction enzyme cutting sites, genes), regardless of inheritance. Distance is measured in base pairs. For the human genome, the lowest resolution physical map is the banding patterns on the 23 different chromosomes; the highest resolution map would be the complete nucleotide sequence of the chromosomes.

Plasmid: Autonomously replicating, extrachromosomal circular DNA molecules, distinct from the normal bacterial genome and nonessential for cell survival under nonselective conditions. Some plasmids are capable of integrating into the host genome. A number of artificially constructed plasmids are used as cloning vectors.

Plate Number (N): A value that describes how good a column is in keeping sample bands narrow. Columns with large plate numbers give narrow bands; long columns packed with small particles give the highest plate numbers.

Point mutation: A mutation in a single base pair within a DNA molecule. See Single nucleotide polymorphism or SNP.

Polar: Compounds, solvents or surfaces that either dissolve in water or prefer water to non-polar organic solvents such as acetonitrile.

Polygenic disorders: Genetic disorders resulting from the combined action of alleles of more than one gene (e. g., heart disease, diabetes, and some cancers). Although such disorders are inherited, they depend on the simultaneous presence of several alleles; thus the hereditary patterns are usually more complex than those of single gene disorders.

Polymerase chain reaction (PCR): A method for amplifying a DNA base sequence using a heat stable polymerase and two primers (usually about 20-25 bases), one complementary to the + strand at one end of the sequence to be amplified and the other complementary to the – strand at the other end. PCR also can be used to detect the existence of the defined sequence in a DNA sample.

Polymerase, DNA or RNA: Enzymes that catalyze the synthesis of nucleic acids on preexisting nucleic acid templates, assembling RNA from ribonucleotides or DNA from deoxyribonucleotides.

Polymorphism: Difference in DNA sequence among individuals. Genetic variations occurring in more than 1% of a population would be considered useful polymorphisms for genetic linkage analysis. Compare mutation.

Pore: Refers to small passage ways or tunnels that crisscross or "honeycomb" a particle. The term is typically used in reference to the small particles which make up the stationary phase of a column. Additionally, the term may also refer to the passage ways in a frit or filtering surface.

Preparative column: A large diameter column used for recovering purified materials by chromatographic separation; can be contrasted with analytical columns which are used for analysis and sometimes recovery of fragments.

Primer: Short preexisting polynucleotide chain to which new deoxyribonucleotides can be added by DNA polymerase.

Probe: Single-stranded DNA or RNA molecules of specific base sequence that are used to detect the complementary base sequence by hybridization.

Prokaryote: Cell or organism lacking a membrane- bound, structurally discrete nucleus and other subcellular compartments. Bacteria are prokaryotes. Compare eukaryote. See chromosomes.

Promoter: A site on DNA to which RNA polymerase will bind and initiate transcription.

Proof-reading: Mechanism for correction of errors made during synthesis of nucleic acids or polypeptides by scrutiny of the products <u>after</u> the nucleotides or amino acids have already been incorporated.

Protein: A large molecule composed of one or more chains of amino acids in a specific order; the order is ultimately determined by the base sequence of nucleotides in the gene coding for the protein. Proteins are required for the structure, function, and regulation of the body's cells, tissues, and organs, and each protein has unique functions. Examples are hormones, enzymes, and antibodies.

Purine: A nitrogen containing basic compound having two fused rings that occurs in nucleic acids. The purines in DNA and RNA are adenine and guanine.

Pyrimidine: A nitrogen containing, single ring, basic compound that occurs in nucleic acids. The pyrimidines in DNA are cytosine and thymine and in RNA are cytosine and uracil.

Qualitative analysis: The determination of which fragments are present in a sample. This is usually done by comparing retention times for bands in the unknown sample fragments with retention times for standard fragments suspected to be present in the sample.

Quantitative analysis: The analysis of a sample to determine the concentrations of the fragments in the sample. Quantitative analysis is usually based on the size of the bands in the sample chromatogram (peak height or area) compared to band size of a sample with a known concentration (the standard).

Recombinant clones: Clones containing recombinant DNA molecules. See recombinant DNA technologies.

Recombinant DNA technologies: Procedures used to join together DNA segments in a cell free system (an environment outside a cell or organism). Under appropriate conditions, a recombinant DNA molecule can enter a cell and replicate there, either autonomously or after it has become integrated into a cellular chromosome.

Recombination: The process by which progeny derive a combination of genes different from that of either parent. In higher organisms, this can occur by crossing over.

Recovery: A term describing the amount of a sample that is actually collected from a column relative to the amount originally introduced into the system, typically described as a "percent."

Regeneration: A process in which the packing inside the column is restored to ist beginning state following a gradient elution. It can also refer to the process of bringing back any column to its original state, usually following temporary damage to the bonded phase.

Regulatory regions or sequences: A DNA base sequence that controls gene expression.

Resolution factor: Defines how well separated two adjacent bands are. Larger values of resolution factor mean better separation.

Retention time: The time between sample injection (time zero) and the appearance of the band maximum; when all conditions are held constant, the retention time for a given band (or compound) remains constant.

Retention Volume: Simply stated, it is the volume of eluent required to elute a substance from the column.

Restriction enzyme or endonuclease: A protein that recognizes specific, short nucleotide sequences and cuts DNA at those sites. Bacteria contain over 400 such enzymes that recognize and cut over 100 different DNA sequences. See restriction enzyme cutting site.

Restriction enzyme cutting site: A specific nucleotide sequence of DNA at which a particular restriction enzyme cuts the DNA. Some sites occur frequently in DNA (e. g., every several hundred base pairs), others much less frequently (rare- cutter; e. g., every 10,000 base pairs).

Restriction fragment length polymorphism (RFLP): Variation between individuals in DNA fragment sizes cut by specific restriction enzymes; polymorphic sequences that result in RFLPs are used as markers on both physical maps and genetic linkage maps. RFLPs are usually caused by mutation at a cutting site preventing cutting. See marker.

Reverse phase chromatography: The most common form of HPLC, in which water-based/polar mobile phases are used with column packings that have a non polar or hydrophobic surface.

Reverse transcription: The process of transcribing RNA into cDNA.

Reverse transcriptase: An enzyme that catalyzes the RNA-dependent polymerization of DNA. This enzymatic activity is found in retroviruses.

RFLP: See restriction fragment length polymorphism.

Ribonucleic acid (RNA): A chemical found in the nucleus and cytoplasm of cells; it plays an important role in protein synthesis and other chemical activities of the cell. The structure of RNA is similar to that of DNA. There are several classes of RNA molecules, including messenger RNA, transfer RNA, ribosomal RNA, and other small RNAs, each serving a different purpose.

Ribonucleotides: See nucleotide.

Ribosomal RNA (rRNA): A class of RNA found in the ribosomes of cells.

Ribosomes: Supramolecular complex composed of specialized ribosomal RNA and protein; site of protein synthesis. See ribonucleic acid (RNA).

RNA: See ribonucleic acid.

RNA polymerase: Enzymatic activity responsible for DNA-dependent synthesis of RNA. In prokaryotes there is only one RNA polymerase. In eukaryotes, there are three, each of which transcribes a different group of genes.

rRNA (ribosomal RNA): RNAs which make up the scaffold upon which the ribosomal proteins are assembled to create the ribosomal subunits.

Sample capacity: The amount of sample that can be introduced onto a column before overload occurs. Column overload is defined as a condition where the efficiency of that column goes below 90 percent of its normal value. It manifests as broad poorly resolved peaks.

Sampling rate: A term describing how frequently the detector generates a reading of the flow cell. Too fast a rate can generate more data than necessary and too slow a rate may allow a narrow band to be missed.

Sensitivity: Can refer to detector sensitivity, which is the ability of the detector to produce intense bands or peaks for a given amount of sample. Sensitivity is usually expressed as the amount of signal for a given mass or concentration of sample. Sensitivity can also refer to the ability of the method to detect a sample if present. In the case of mutation analysis, a 95 % sensitivity means the method will detect 95 % of the mutations present in the sample. This term must be used carefully to avoid confusion.

Separation Factor: Defined as the relative retention measured for two adjacent peaks, either through measurement of the difference between the centers of the peaks or between the tops of the peaks. Also referred to as selectivity.

Sequence: See base sequence.

Sequence tagged site (STS): Short (200 to 500 base pairs) DNA sequence that has a single occurrence in the human genome and whose location and base sequence are known. Detectable by polymerase chain reaction, STSs are useful for localizing and orienting the mapping and sequence data reported from many different laboratories and serve as landmarks on the developing physical map of the human genome. Expressed sequence tags (ESTs) are STSs derived from cDNAs.

Sequencing: Determination of the order of nucleotides (base sequences) in a DNA or RNA molecule or the order of amino acids in a protein.

Sex chromosomes: The X and Y chromosomes in human beings that determine the sex of an individual. Females have two X chromosomes in diploid cells; males have an X and a Y chromosome. The sex chromosomes comprise the 23rd chromosome pair in a karyotype.

Shotgun method: Cloning of DNA fragments randomly generated from a genome.

Single gene disorder: Hereditary disorder caused by a mutant allele of a single gene (e. g., Duchenne muscular dystrophy, retinoblastoma, sickle cell disease). Compare polygenic disorders.

Single nucleotide polymorphism: Polymorphism where a single nucleotide is different from the corresponding wild type DNA sequence. See SNP.

Solute: A sample fragment that is separated using a column.

Solvents: The pure components of the mobile phase (e. g., water, methanol, acetonitrile, etc.). In reverse phase chromatography, water is a weak solvent, so eluents with higher concentrations of water are weaker and give longer retention times for all sample bands. Acetonitrile is a strong solvent; increasing the amount of acetonitrile in the eluent makes it stronger, and sample bands leave the column sooner.

Solvent Strength: The ability of a solvent to elute a particular sample fragment or solute from a column.

Somatic: Refers to non-germline cells. Somatic cells may become terminally differentiated with alterations in their overall genetic compliment, because they are <u>not</u> responsible for passing along the organism's genetic material to the offspring. See germline.

Somatic hypermutation: A very high frequency of mutational events that occur in specific loci, e. g. the genes corresponding to the variable segments of expressed immunoglobulin.

Southern blotting: Transfer by absorption of DNA fragments separated in electrophoretic gels to membrane filters for detection of specific base sequences by radiolabeled complementary probes.

SNP: Single nucleotide polymorphism.

Stainless steel: Any variety of steel alloys designed for corrosion resistance. The different varieties – primarily 316 grade – are used to manufacture high-pressure, chemically-resistant HPLC fittings and tubing.

Standard: A sample with a known concentration or known identity of a fragment.

Stationary phase: The particulate material packed inside the column.

Start codon: That codon at which translation of an mRNA molecule begins. This is always an AUG (encoding methionine) in eukaryotes, and nearly always in prokaryotes. In prokaryotes N-formyl-methionine is used to initiate polypeptide synthesis.

Stop codon: That codon at which translation of an mRNA molecule into a polypeptide is terminated. In the Universal Code this may be: UGA, UAG, or UAA.

Stringency: A measure of the strength of (hydrogen) bonding two complementary strands of DNA. This term is often applied to nucleic acid hybridization conditions to indicate the degree of sequence homology between probe and target sequences.

Strong mobile phase: A strong mobile phase will give shorter retention times for all sample bands; in reverse phase chromatography, acetonitrile is a strong solvents, and water is a weak solvent.

STS: See sequence tagged site.

Tailing: A non Gaussian curve in which the peak "leads out," and is shown by slight tapering on the bottom trailing edge of the peak's trace. Tailing is the opposite of fronting.

Tandem repeat sequences: Multiple copies of the same base sequence on a chromosome; used as a marker in physical mapping.

Teflon® (PTFE, FEP, PFA): A class of chemically inert fluoropolymers used to make fittings, tubing and accessories primarily for low pressure applications. Teflon is a Registered Trademark of the DuPont Co.

Tefzel® (ETFE): A molecular "cousin" to Teflon, this is a chemically resistant fluoropolymer used to make fittings, tubing and accessories primarily for low to moderate pressure applications. Tefzel is a Registered Trademark of the DuPont Co.

Telomere: The ends of chromosomes. These specialized structures are involved in the replication and stability of linear DNA molecules.

Test chromatogram: A chromatogram from the column manufacturer that usually accompanies a new column, showing the separation of a particular sample under some set of experimental conditions. The test chromatogram shows how well the column performs under standard (and generally ideal) conditions.

Theoretical plate: A measure of column efficiency; As the number of theoretical plates increases for a given column type, the greater the resolution that may occur between sample fragments.

Thymine (T): A nitrogenous base, one member of the base pair A–T (adenine- thymine).

Titanium: An alternative to stainless steel when a DNA compatible metal part is desired.

Transcription: The synthesis of an RNA copy from a sequence of DNA (a gene); the first step in gene expression. Compare translation.

Transfer RNA (tRNA): A class of RNA having structures with triplet nucleotide sequences that are complementary to the triplet nucleotide coding sequences of mRNA. The role of tRNAs in protein synthesis is to bring amino acids to the ribosomes, where proteins are assembled according to the genetic code carried by mRNA.

Transformation: A process by which the genetic material carried by an individual cell is altered by incorporation of exogenous DNA into its genome.

Translation: The process in which the genetic code carried by mRNA directs the synthesis of proteins from amino acids. Compare transcription.

tRNA: See transfer RNA.

Uracil: A nitrogenous base normally found in RNA but not DNA; uracil is capable of forming a base pair with adenine.

Ultraviolet/Visible Light (UV/Vis) or UV Detector: The most popular form of detector used in DNA Chromatography. This device typically offers a tunable wavelength feature, allowing operation at any wavelength from 190 nm to near 800 nm. The most popular setting is either 260 nm (or 254 nm for single wavelength detectors.

Union: Used to join 2 pieces of tubing, usually with the same outer diameter.

Vector: See cloning vector.

Virus: A noncellular biological entity that can reproduce only within a host cell. Viruses consist of nucleic acid covered by protein; some animal viruses are also surrounded by membrane. Inside the infected cell, the virus uses the synthetic capability of the host to produce progeny virus.

Void (or column void): A space in the column packing, generally at the inlet of the column, or also a settling of the packing at the column inlet to create a space between the top of the packing and the frit. A void usually makes the column unsuitable for use.

Void volume: The total internal volume of a connection or fluid pathway, also known as dead volume swept volume or hold up volume.

Waste Container: An essential component for any HPLC system, it is a vessel at the end of the system used to collect the detector effluent. The waste container should be located inside of a spill tray to prevent accidental flooding of the laboratory bench or floor.

Weak mobile phase: A weak mobile phase gives longer retention times for sample bands, compared to a strong mobile phase. Water is a weak solvent and acetonitrile is a strong solvent for reverse phase chromatography. Mobile phases with more water will be weaker, and mobile phases with more organic solvent will be stronger.

YAC: See yeast artificial chromosome.

Yeast artificial chromosome (YAC): A vector used to clone DNA fragments (up to 400 kb); it is constructed from the telomeric, centromeric, and replication origin sequences needed for replication in yeast cells. Compare cloning vector, cosmid.

Appendix 2
System Cleaning and Passivation Treatment

A 2.1
Background Information

HPLC separation problems caused by corroding stainless steel surfaces have been reported for the analysis of inorganic ions and proteins. Poor electrochemical detection (detection by oxidation or reduction of the sample peak) has sometimes been shown to be due to metal ion contamination found or released from stainless steel surfaces [1, 2]. While systems free of metal contamination has been needed for some types of HPLC, DNA Chromatography exhibits the highest system requirements needed to be able to reliably separate and detect nucleic acid separations. The surfaces of the system cannot contain metal ions that will either trap the nucleic acids or perform structural or conformational changes to the nucleic acid. Nor can the surfaces release metal ions to travel to another point within the system (the column end inlet frit for example) that will damage the separation [3–6].

Fortunately, it is possible to clean and maintain stainless steel components or other (plastic or metal) materials that contain harmful metal ions [7, 8]. Most manufacturers of HPLC equipment have standard operating procedures on how to clean and passivate their equipment. These original procedures have been further modified to ensure the DNA Chromatography system surfaces are contamination free and stable. Treatment with nitric acid will clean and passivate a stainless steel and make it resistant against further corrosion. The concentrations of nitric acid used for HPLC system passivation vary from 3 to 13 M; however, treatment with 8 M nitric acid seems to be sufficiently rapid and effective for cleaning a DNA Chromatograph.

In the cleaning mechanism, nitric acid will remove existing metal ion contamination and some organic contaminants. For example, nitric acid is used to clean organic and metal material on quartz detector windows in UV detectors and it is used to remove surface corrosion of the tubing of the system. This does not mean that this cleaning procedure should be used to clean incompatible components. For example a (rusty prone) stainless steel frit is not a suitable component for a DNA Chromatograph. And a system that has worked well for protein separations for example will not necessarily work for DNA separations. Stainless steel is an excellent material for HPLC and is used in systems where high structural strength or reliability is required. However, stainless steel frits (that contain very high surface area and may easily corrode) should be replaced with a titanium or polymer frit material rather than relying on system cleaning to maintain performance. Even then, titanium or polymers may contain metal ion contamination that must be removed.

In the passivation mechanism, nitric acid oxidizes the metal surface by producing a very thin oxide layer, which protects the metal from further attack by oxygen, acid, and other influences. Nitric acid will also preferentially remove iron from the surface (or sharp edges prone to corrosion) effectively leaving chromium and nickel metal at relatively higher concentrations at the surface. These metals are less likely than iron to undergo further corrosion.

In the procedure described here, a treatment with a chelating reagent is included to ensure complete cleaning and passivation. System treatment should be performed once every 6 months or whenever necessary due to deteriorating column performance (broad, inconsistent or missing peaks). It is a good idea to change piston seals, injection port seals and flow path filter after the passivation. Even a new column could show deteriorating performance. In fact, if performance degradation of a new column occurs quickly, it is a strong sign that the system needs cleaning.

CAUTION: Nitric acid can be dangerous. Always wear safety glasses and gloves. The nitric acid effluent should be collected in a separate (empty) waste container. Do not mix concentrated nitric acid with organic solvents of any kind.

The operator performing the passivation must be familiar with the basic operation of the DNA Chromatographic system and must know how to remove and install the injection port and seal and the injection needle. Purging and rinsing may have to be done in manual mode with 50% A and 50% B to passivate both reservoir flow paths. The column and inline degassers should be removed from the flow path to prevent harming these components. The detector should also be removed unless there is a need to clean this component. The recommended flow rate for cleaning and rinsing is 2 ml/min. Do not close the door to the column heater or turn off the heater. Passivation should be performed at room temperature.

An experiment was performed to show the effect of passivation of a new system that had not yet undergone final cleaning and conditioning. The identical column was used for all four chromatograms. Figures A2.1 and A2.2 show how a contaminated system might perform for a size standard and a mutation standard. In these cases, the peak pattern is unusual and the resolution of some peaks is poor. In other cases (discussed in Chapter 3, the peaks may split into doublets or even disappear. Figures A2.3 and A2.4 show how the separation should look in a clean system. In these cases, the peak patterns are normal (as compared with the standard chromatograms included with the new column) and the peak resolution is excellent. This cleaning and passivation procedure should not to be considered as an exotic rescue attempt but rather as a simple way to prevent premature system failure.

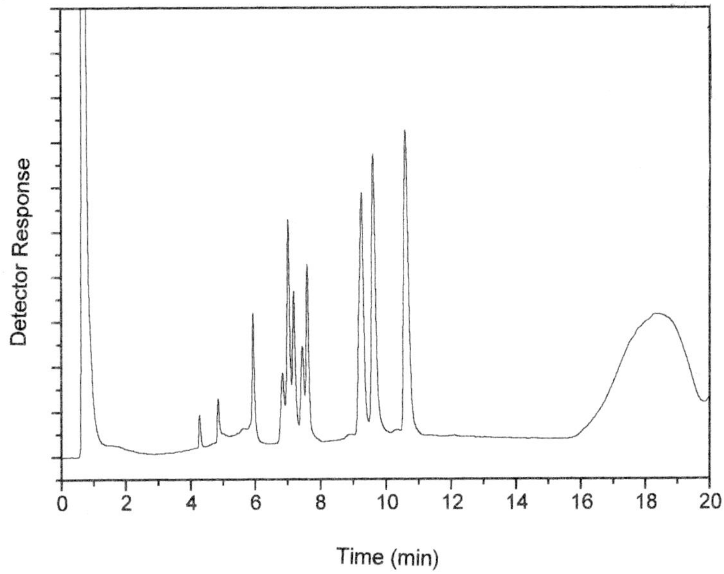

Figure A2.1. Separation of pUC 18 *Hae* III digest size standard on a new DNA Chromatogrampic System before system treatment. Separation conditions are specified as standard test conditions for performing a size based separation for a new DNASep® column. From Transgenomic, Inc. with permission.

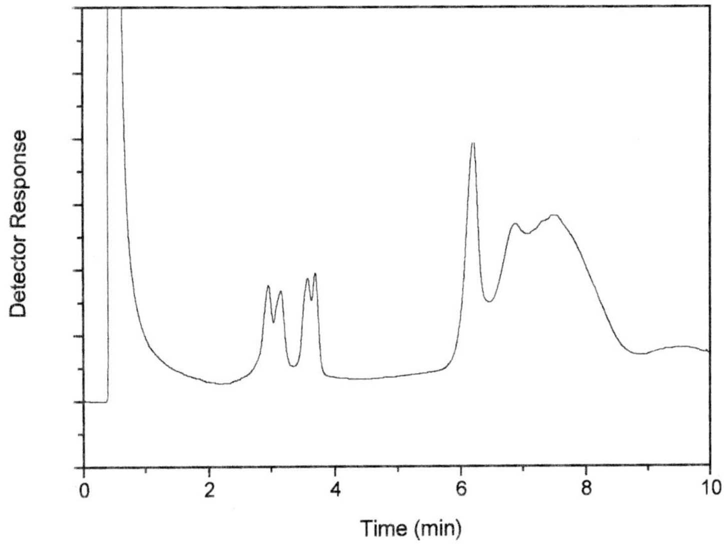

Figure A2.2. Separation of Dys271 mutation standard on new DNA Chromatgraphic System before system treatment. Separation conditions are specified as standard test conditions for performing a size based separation for a new DNASep® column. From Transgenomic, Inc. with permission.

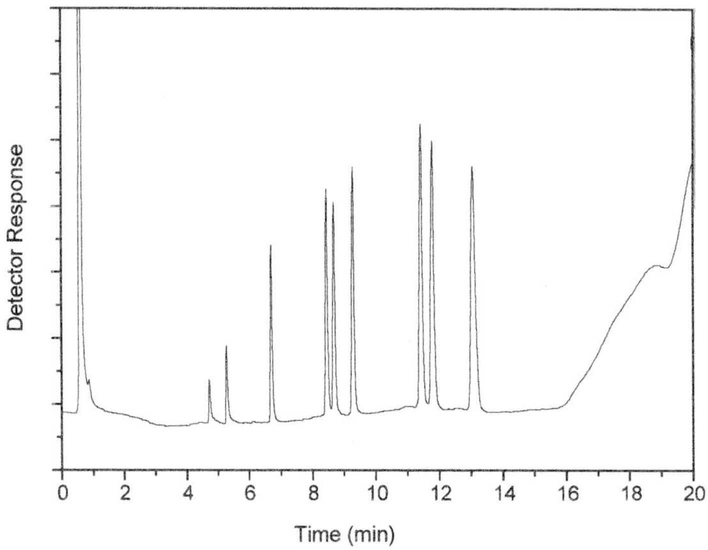

Figure A2.3. Separation of pUC 18 *Hae* III digest size standard on a new DNA Chromatogrampic System after system treatment. Separation conditions are specified as standard test conditions for performing a size based separation for a new DNASep® column. From Transgenomic, Inc. with permission.

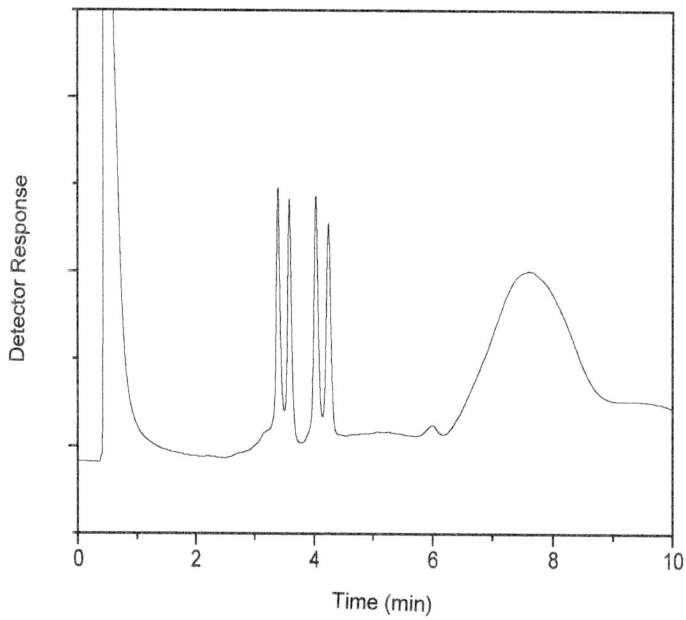

Figure A2.4. Separation of Dys271 mutation standard on new DNA Chromatgraphic System after system treatment. Separation conditions are specified as standard test conditions for performing a size based separation for a new DNASep® column. From Transgenomic, Inc. with permission.

A2.2
Reagents

1. DI Water
2. 8 M Nitric acid (approx. 35 % w/w)
   Preparation: In 250 ml media bottle, add slowly 70 ml of concentrated nitric acid (HNO_3, 69.5%, d = 1.4 g/mL) to 100 ml of DI water. The solution will get warm; wait until the acid has reached room temperature before using.
3. Ethylenediamine tetraacetic acid, tetrasodium salt (tetrasodium EDTA), 50 mM (Na_4EDTA)
   Preparation: Dissolve 22.6 g of tetrasodium EDTA (FW = 452.24) in about 500–700 ml of DI water (1 L volumetric flask). After all the solids have dissolved dilute with DI water to the 1 L mark.

A2.3
Preparation of the System

1. Replace the column with a PEEK™ union.
2. Bypass detector flow cell, with Teflon tubing (0.010 or 0.020"ID), from preheat coil to a new waste container of appropriate length.
3. Rinse entire system with distilled water (all channel reservoirs): purge for 5 minutes and flush for another 30 min at 2 ml/min. Check for any leaks! Make sure system is without leaks. Do not close door to column oven or turn off oven.

A2.4
Passivation of System

1. Remove the white solvent inlet filter caps from the solvent lines of all channel reservoirs.
2. Fill 8 M nitric acid in appropriate glass bottle (Erlenmeyer flask, media bottle) and insert tubing ends of all channel reservoirs into the acid.
3. Rinse all channel reservoirs with 8 M nitric acid: purge for 5 min; rinse system for another 15 min at 2 ml/min.
4. Replace nitric acid with DI water. Rinse all channel reservoirs DI water: purge for 10 min and flush for 90 minutes at 2 ml/min. During this time change the DI water at least three times.
5. Check the pH of the effluent with pH paper. Keep rinsing with water until pH >5.
6. Replace the DI water with a 50 mM solution of tetrasodium EDTA (Na_4EDTA). Purge for 5 min and rinse system for 30 min at 2 ml/min. Caution: If nitric acid is not rinsed thoroughly, the EDTA could precipitate plugging the system.

A2.5
Equilibration of System

1. Install new inlet filter caps.
2. Replace Na_4EDTA solution with water, purge for 5 min and rinse for 15 minutes.
3. Replace water with fresh eluents (A and B), purge for 5 min and rinse for at least 2 hours at 2 ml/min.
4. Install column, equilibrate column with at least one gradient run and test system with pUC 18 *Hae* III digest and Dys271 mutation standard.

A2.6
Passivation of Injection Port and Injection Needle

It is a good idea to passivate injection port and needle at the same time of system passivation. The same reagents are used for this process. In addition a small glass beaker (100 ml, tall form) and a sonicator (dental cleaner) are necessary.

1. Remove injection port and needle. Remove injection port seal from the port
2. Rinse port and needle with DI water
3. Put needle and port into beaker.
4. Fill beaker with 8 M nitric acid until needle and port are covered
5. Place beaker in sonicator and sonicate for 5 min (under fume hood)
6. Decant nitric acid into recycle bottle. Acid can be reused for passivation.
7. Rinse needle and port several times with DI water.
8. Sonicate needle and port in fresh DI water for 5 minutes.
9. Check pH. If $pH<4$ replace DI water and sonicate 5 minutes. Repeat until $pH>5$.
10. Replace water with 50 mM Na_4EDTA solution and sonicate for 5 minutes.
11. Rinse with DI water one more time.
12. Install new injection port seal.
13. Re install needle and port.

References

1. C. Duda and C. S. Brunlett, Amperometric Techniques, in Handbook Instrumental Tech. Anal. Chem. Bioanalytical Systems, F. A. Settler ed., Prentice Hall: Upper Saddle River, NJ, pp 691–708, 1997
2. K. E. Collins, C. H. Collins and C. A. Bertran, Stainless steel surfaces in LC systems, Part I – corrosion and erosion, LC/GC Mag., 18, 600, 2000.
3. D. T. Gjerde, R. M. Haefele, and D. W. Togami, Method for performing polynucleotide separations using liquid chromatography, U. S. Pat. 6,017,457, 2000.
4. D. T. Gjerde, R. M. Haefele, and D. W. Togami, System and method for performing polynucleotide separations using liquid chromatography, U. S. Pat. 5,772,889, 1998.
5. D. T. Gjerde, R. M. Haefele, and D. W. Togami, Apparatus for performing polynucleotide separations using liquid chromatography, U. S. Pat. 6,030,527, 2000.
6. D. T. Gjerde, R. M. Haefele, and D. W. Togami, Liquid chromatography systems for performing polynucleotide separations, U. S. Pat. 5,997,742, 1999.
7. K. E. Collins, C. H. Collins and C. A. Bertran, Stainless steel surfaces in LC systems, Part II – passivation and practical recommendations, LC/GC Mag., 18, 688, 2000.
8. R. Shoup and M. Bogdan, Passivation of liquid chromatography components, LC/GC Mag., 7, 742, 1989.

Appendix 3
Frequently Asked DHPLC Questions

Joanne Walter

A3.1
What are the various methods for DHPLC temperature selection?

There are two major methods to determine the temperature needed to perform DHPLC analysis. Traditionally, a melting curve of homoduplex retention time vs. oven temperature can be constructed [1]. Alternatively, software analysis of the fragment sequence can be used to predict the optimum DHPLC temperature(s) [2, 3]. Both methods predict the average melting temperature (T_M) of the fragment although additional information can be gained from the software approach.

In order to construct a melting curve, it is necessary to carry out repeated injections of the wild type or reference sample starting at an oven temperature of 50 °C and ending at 70 °C. A shallow eluent gradient employing a wide acetonitrile concentration range is used to ensure that both double-stranded and single-stranded DNA retention times can be measured (none of the fragment peaks elute with the injection peak or the wash off peak). If no prior information about the melting point and melting range of the fragment is known then regular 2 °C intervals should be chosen. The interval and range of temperature can be adjusted if some information is known about the fragment melting such as GC content. A melting curve may have to be constructed if a new column is being used, an oven has been recalibrated, or an old column is suspect and must be tested.

The melting curve is constructed by plotting retention time on the y-axis and temperature on the x-axis. The resulting curve can be used to determine the T_M of the fragment by measuring the retention time of double-stranded DNA (marked with arrow A in Figure A3.1) and single-stranded DNA (marked with arrow B), measuring the midpoint between the two retention time and then reading the corresponding midpoint temperature (T_M) from the x axis (marked by the dotted line). The ideal DHPLC temperature is then defined as being 1 °C lower than the midpoint (T_M-1). T_M may also be defined as the point where the DNA is 75 % helical (25 % melted). The recommended analysis temperature is usually rounded off to the nearest whole degree.

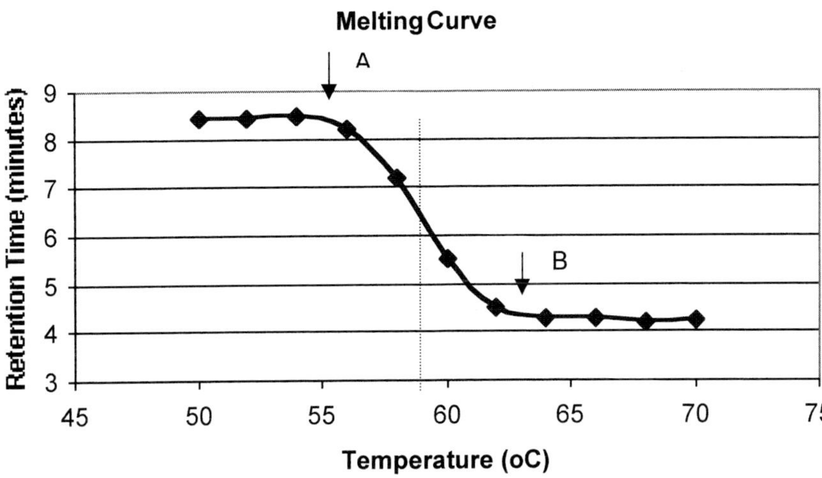

Figure A3.1. This is a typical "S" shaped melting curve with an approximate T_M of 59 °C making temperature of analysis for DHPLC to be 58 °C. See Chapter 4 for a further description of how the plot is measured.

The melting curve approach to temperature determination is accurate but time consuming and is not ideal when large numbers sample with different sequences are to be analyzed. The alternative approach is to import the fragment sequence to be analyzed into either of two software based DHPLC temperature prediction programs [2, 3]. Both programs use sequence interpretation algorithms to interpret the nucleotide composition and the ordering of bases within a sequence. This enables determination of the optimum temperature in a matter of minutes compared to the 2–3 hours it can take for a conventional melting curve temperature prediction. The programs are the WAVEMAKER® program (Transgenomic, Inc.) and a program available from the Stanford University web site. A major strength of DHPLC is that mutations are detected over a wide range of temperatures: usually 2–3 °C but some references quote as high a 6 °C range [2, 4]. Thus, any software based method of temperature prediction only needs to be within this temperature range to enable the detection of mutations. Both programs meet this specification.

A3.2
Does having the optimum oven temperature mean that I will get the optimum resolution of the heteroduplex and homoduplex species?

The surprising answer is no. Remember the goal is to be able to detect all mutations regardless of where they are on the fragment. Since the optimum temperature for one mutation might be different than the optimum temperature for another mutation a compromise must be made. This does not mean that mutations separated at a less than optimum temperature cannot be detected. It only means that the highest peak resolution will not be achieved. By following the guidelines described in this section and in the book all mutations can be detected.

The situation changes when a group of samples is being analyzed for only 1 known mutation. Then the temperature may be adjusted to improve the resolution to its optimum value.

A3.3
I do not have the complete sequence of my fragment but I want to scan for mutations. Is this still possible by DHPLC?

In certain cases, only the primer sequence used to amplify the fragment is known but the rest of the fragment sequence is unknown. It is possible to analyze these samples provided a careful melting curve is constructed (see the answer to Question 1). The temperature titration should be performed with the temperatures closely spaced together (usually 1 °C) to give the best chance of not only measuring the average T_M but also to determine the T_M of any melting domains that might be present. Figure A3.2 shows a melting curve from a fragment where there are multiple temperature domains present. Instead of performing DHPLC analysis just at the average $T_M - 1$ (55 °C), two temperatures should be used to find all mutations in the entire fragment.

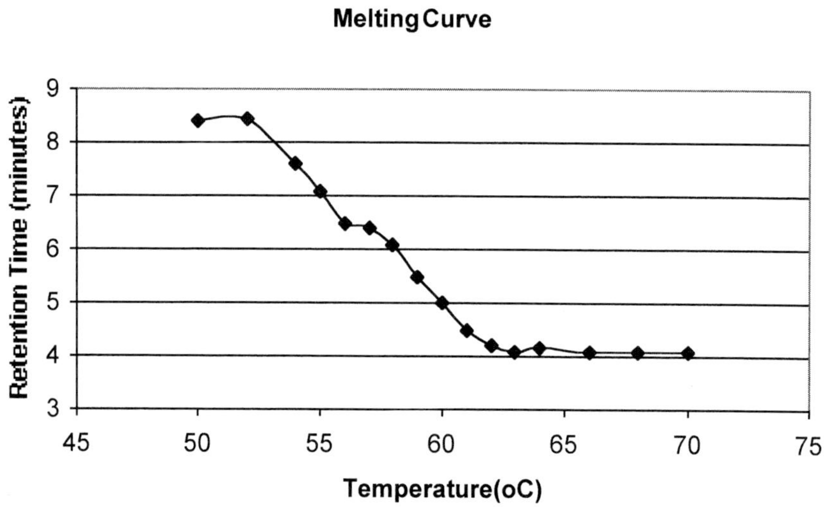

Figure A3.2. The T_M from the melting curve is approximately 56 °C and therefore either 55 °C (T_M−1) would be chosen as the DHPLC analysis temperature. However, because there are two inflection points indicating more than one domain, it may be preferential to choose 53–54 °C for the first domain and 58–59 °C for the second domain. This would ensure that mutations throughout the entire fragment are located.

A3.4
Will DHPLC detect both heterozygous and homozygous mutations?

Formation of a heteroduplex is necessary to detect a mutation by DHPLC. Heteroduplices are formed by heating the sample to 95 °C for 5 minutes to completely denature the DNA and then slowly cooling the sample to allow the strands to re-hybridize. As the DNA reanneals, a portion of the sample will form perfect (homodulplex) Watson – Crick pairing along the whole length of the fragment. A heteroduplex is formed when a wild type strand is paired with a mutant strand creating a mismatch at one or more base sites.

In a heterozygous mutant sample only one of the chromosomes is mutant so heteroduplex and homoduplex species are formed after PCR amplification during the reannealing process. In a homozygous mutation sample both chromosomes are mutant and heteroduplex species cannot form because all strands are complementary. The PCR sample must be first premixed with a wild type or reference sample and the reannealed to form the heterduplex (and homoduplex) species.

Adding wild type reference material to the samples will detect all mutations regardless of type. Mutation verification and classification of whether they are heterozygous or homozygous can be done by sequencing.

A3.5
I have a sample population to scan for mutations and need to be certain that I find all mutations present. What are the factors affecting the accuracy of DHPLC and how should I approach the problem?

Although it is unlikely that any one technique is capable of finding all mutations in any given sequence, DHPLC has been found to consistently attain high levels of accuracy [5, 6]. In order to ensure that the maximum number of mutations are located within a fragment, it is necessary to use the software approach to carry out an analysis of the fragment melting domains. A melting domain is a region within the fragment where all of the bases melt at the same temperature. A single fragment may have many different domains that melt at different temperatures and this will affect whether a single temperature or several oven temperatures are needed for DHPLC analysis.

To start, the sequence of the fragment is imported into the program and the fragment is analyzed for the average T_M and the T_Ms of the various temperature domains (if more than 1 is present). If temperature domains in a fragment that make up the average temperature results are very different from each other, there is a chance of missing mutations when analyzing at only the average T_M 1 value [7]. As stated in the answer to Question 1, mutations are easily visible within a 2–3 °C window of temperature and only one temperature will be needed if the melting domains fall within that temperature range [3]. If the T_Ms of domains are more than 4 °C apart it is necessary to use the T_M−1 temperatures for each of the individual domains. Mutations in higher melting domains are hard to detect than lower melting domains more care should be taken in these cases.

Both software programs predict the melting domains of the sequences. The Stanford program predicts the general analysis temperature and then prints out the temperatures of all the domains. This can be exhaustive as it may define a domain as only a few base pairs. The WAVEMAKER® program provides interactive graphics that allow the user to install his own rules surrounding temperature prediction and what is defined as a domain.

Temperature prediction is obviously the main factor affecting the accuracy of DHPLC. Nevertheless, the ability of DHPLC to detect mutations over a range of temperatures is the reason for its ability to detect mutations with high accuracy. The position of mutation (whether it is close to the primer site or in middle of the fragment) has been shown to be irrelevant to the accuracy of the technique (provided the melting domain analysis has been carried out). The type of mutation (base change) also does not affect greatly on whether the mutation can be detected (although in some limited cases DHPLC has been used to determine the type of mutation base change). The mutation detection method is largely independent of fragment size (see the next question).

The fidelity of the polymerase used for PCR can have a large effect on detection. The misincorporations of the PCR enzyme (sometimes call PCR induced mutations) are visualized as low level, random mutations characteristically seen on the chromatogram as a bump of varying size before the main homoduplex peak. If the level of this preceding peak is high, it can obscure small changes in sample peak pattern and may cause a mutation to be missed. Many laboratories are able to perform DHPLC analysis routinely without the use of a proof reading polymerase. However, if PCR induced errors are a problem, careful PCR (careful measurement of reagents and careful use of the thermocycler) and especially changing to a proof-reading PCR enzyme has been found to made dramatic improvements in the detection of mutations.

A3.6
What are the minimum and maximum fragment sizes I can analyze by DHPLC?

The lower size limit for the technique has been reported to be in the 100–150 bp range whereas mutations have been located in fragments as large as 1500 bp [8]. Unless there is some experimental constraint such, a size range of 200–800 bp is reasonable with the majority of researchers analyzing fragment sizes between 300–700 bp. But this fragment size is not always possible. For example, the amplification may be occurring from archival samples (i. e. paraffin-embedded tissues). In this instance, one may wish to amplify only a short stretch of sequence so as to reduce the deleterious effects of genomic DNA degradation.

Smaller fragment samples may contain primer dimer impurities formed in the PCR process or single-stranded fragments with primer overhangs formed during the reannealing step. These impurities may coelute with the small fragments leading to potential detection complications. Also, the homoduplex and heteroduplex species of small fragments may both melt together making it difficult to distinguish the complexes. Adding a GC clamp both lengthens the fragment and establishes a higher, more stable temperature melting domain making detection of the mutation in the lower melting domain easier. A GC clamp may also help to detect mutations in larger fragments for the same reason.

At the upper fragment size (1500 bp) the resolving power of the column is decreased making resolution of the heteroduplex and homoduplex more difficult. Furthermore, a single base change in a larger fragment does not have the thermodynamic impact than in a smaller fragment. The heteroduplex and homoduplex melting temperatures are not so different making it more difficult to detect the heteroduplex.

A3.7
I need to screen a large number of samples. What is the quickest way to do this?

One approach is simply to analyze samples on a 24-hour basis. Conventional DHPLC instrumentation will allow between 150-200 samples per 24-hour period. Recent instrumentation advances may increase sample throughput to 300-350 samples per day.

Another approach is to pool samples of the same fragment type to enable the simultaneous processing of many samples in the same chromatographic run. What should define the pool size? The limits of detection are a key criteria. Single mutations in pool sizes of 4 –10 samples are usually detectable.

Another factor in pooling is the frequency of mutations expected in the sample population being analysed [13]. If the polymorphisms in the fragments are common, the pool size should be small so that every pool does not give a positive result. All samples within a positive pool must be analyzed to determine which sample or samples are positive. In cases where many of the 10 sample pools show a positive signal, a pool size of 4 may be more appropriate.

A3.8
I want to create a general SNP map but do not need to find every mutation. What is the best strategy?

The problem is to analyze a large number of sequence types while maintaining a reasonable accuracy for finding their mutations variation. To increase speed of analysis usually only 1 oven temperature should be used unless there is an extreme difference in fragment domain structure.

First, method prediction should be by one of the automated software programs and the temperature chosen should be at the average melting temperature of the fragment. If a maximum fragment size of 400-500 bp is defined for fragment amplification then the efficiency of finding mutations at 1 temperature of analysis should be approximately 90 %. But it should be recognized that no discoveries will be made in regions of the fragment where melting deviates greatly from the average melting temperature.

The optimum number of samples should be determined. As the number of samples is increased, the number of different mutations that will be discovered will also increase. When searching for common polymorphisms, it is normal to analyse 16-20 samples while expecting to find 1 or 2 positive results. However, if rare mutations are to be discovered, then the number of samples to be analyzed is dependent on the rarity of mutation being sought. It is common to analyze 100-200 samples with each set of conditions [9].

A3.9
What will happen if I have more than one mutation in a fragment? Does each mutation or combination of mutations give a unique chromatographic pattern?

Any mutation present in a fragment will be detected by DHPLC as a change in its chromatographic pattern from the wild type or reference sample. The presence of a mutation is indicated, but the type and exact location of the mutation is not known. The presence of an additional mutation will make further changes in the melting characteristics of the fragment and hence change the chromatographic pattern.

The chromatographic pattern is likely to say something about the mutation. A complex pattern may mean that more than one mutation is present (although multiple mutations are less likely). A well resolved 4 peak pattern is likely to indicate the mutation is present a low melting domain in the fragment. A poorly resolved pattern may mean that the mutation is in a high melting domain or may mean that the oven is not at the optimum temperature for the mutation detected.

So it is true. All things being equal, a particular mutation (or combination of mutations) has its own characteristic pattern. This is assuming that there are no changes in the column condition or oven temperature. But this is not the whole story (see the next question).

A3.10
What about the converse? Does a particular chromatographic pattern indicate a particular mutation?

A particular chromatographic pattern may be unique but there is not enough fine structure to the pattern to be certain that a particular pattern indicates a particular mutation. A pattern that looks the same can indicate many different types of mutations. Confirmation of the mutation by sequencing is necessary to positively identify it. As stated before, the pattern is also related to the condition of the column, eluent and instrument. Therefore the pattern may change slightly from laboratory to laboratory or from day to day.

If the fragment sequence being measured is known to contain only a few different mutations and the patterns are different then the peak pattern may be used to identify the mutation. Under controlled circumstances, this property has been used in diagnostic work to identify particular mutations (see Question 12). This is assuming that no undiscovered mutations are present. A standard for each mutation type should be run along with the samples to help with identification and to ensure the instrument is working properly.

A3.11
I have a biological system where it is likely that a mutation if present can have a heterduplex concentration of less than 50%. What is my best approach to this problem?

DHPLC has been shown to be capable of locating mutations present at low levels in a sample population [10]. Mutations in tumor populations and heteroplasmy in mitochondria [11] are two examples where DHPLC has performed well due to its ability to measure low concentrations of mutations. Temperature prediction will be the same as for conventional DHPLC, but particular attention should be paid to achieving good resolution of the heteroduplex from the homoduplex by making sure the fragment sequence being scanned is in a low temperature domain relative to the rest of the fragment. Also, is important to consider the polymerase fidelity. The misincorporation rate of the polymerase can be critical because the characteristic misincorporation peak eluting prior to the main peak may mask a small change that would normally indicate a low-level mutation. A proof reading enzyme that is compatible with the column technology (one that won't plug the column) should be chosen to achieve high sensitivity.

Will fluorescence detection improve sensitivity? Careful use of terms is needed. The use of fluorescence will allow lower amounts of DNA to be detected [12]. However, this detection will not improve the resolution between hetero- and homoduplices. If the resolution of heteroduplex and homoduplex is poor, small amounts of both of these complexes can be seen, but the ability to detect smaller amounts of heteroduplex relative to homoduplex species is unaffected. Good resolution is needed to improve the detection of the heteroduplex and increase the sensitivity of detecting a mutation. Low misincorporation rates (low background signal) are also needed to increase sensitivity when fluorescence detection is used [Ref. [11] and Chapter 4].

Also to be considered is the method of mutation verification after DHPLC detection. Conventional sequencing will not detect the presence of a sequence change if it is present at less than 30% in a population. Obviously the DNA could be cloned and then a number of colonies picked for sequencing but this is extremely laborious. A better method is to collect the heteroduplex from the column, either manually for with an automated fragment collector. The mutant DNA population is enriched and conventional sequencing can now be used to identify the sequence.

A3.12
Is it possible to use DHPLC in a diagnostic setting?

It is possible to use DHPLC in a diagnostic setting [14]. The mutation chromatographic pattern for a particular fragment and set of conditions can be indicative of a particular mutation. However, it is possible that more than one mutation within the same fragment can lead to the same chromatographic pattern and hence lead to an incorrect diagnosis [8]. The mutation giving the same pattern may be known or unknown. This would lead to false positive results but would not cause the more serious false negative results.

In many cases, there are a limited number of mutations and the sample load per day may be low. A versatile method may be needed so that the type of diagnostic test may be changed easily and rapidly. In these cases, DHPLC is an excellent method. If the location of the mutation is not known as is the case in many cancers, then the ability of DHPLC to measure unknown mutations makes it an excellent choice.

For larger sample loads, it may be preferable to design a genotyping assay. Mini sequencing or primer extension assays can be accomplished in an automated way using the DNA Chromatography instrumentation and column technology [15].

References

1. W. Liu, D. I. Smith, K. J. Rechtzigel, S. N. Thibodeau, and D. C. James, Denaturing high performance liquid chromatography (DHPLC) used in the detection of germline and somatic mutations. Nucleic Acids Res., 26, 1396, 1998.
2. W. Xiao and P. J. Oefner, Denaturing high performance liquid chromatography: a review. Human Mutation, 17, 439, 2001.
3. J. I. Spiegelman, N. M. Michael and P. J. Oefner, High accuracy DNA sequence variation screening by DHPLC, Biotechniques, 29, 1084, 2000.
4. A. C. Jones, J. R. Sampson, B. Hoogendoorn, C. Cohen, and J. P. Cheadle, Application and evaluation of denaturing HPLC for molecular genetic analysis in tuberous sclerosis, Human Genetics, 106, 663, 2000.
5. V. N. Kristensen, D. Kelefiotis, T. Kristensen and A. Borresen-Dale, High-throughput methods for detection of genetic variation, Biotechniques, 30, 318, 2001.
6. A. Erlandson, H. Stibler, B. Kristiansson, J. Wahlstrom, and T. Martinsson, Denaturing high performance liquid chromatography is a suitable method for PMM2 mutation screening in carbohydrate-deficient glycoprotein syndrome type 1A patients, Genetic Testing, 3, 293, 2000.
7. M. Cargill, D. Altshuler, J. Ireland, P. Sklar, K. Ardlie, N. Patil, C. R. Lane, E. P. Lim, N. Kalyanaraman, J. Nemesh, L. Ziaugra, L. Friedland, A. Rolfe, J. Warrington, R. Lipshutz, G. Q. Daley and E. S. Lander, Characterization of single-nucleotide polymorphisms in coding regions of human genes, Nature Genetics, 22, 231, 1999.
8. M. C. O'Donovan, P. J. Oefner, S. C. Roberts, J. Austin, B. Hoogendoorn, C. Guy, G. Speight, M. Upadhaya, S. S. Sommer and P. McGuffin, Blind analysis of denaturing high performance liquid chromatography as a tool for mutation detection, Genomics, 52, 44, 1998.
9. C. E. Glatt, J. A. DeYoung, S. Delgado, K. S. Service, K. M. Giacomini, R. H. Edwards, N. Risch, and N. B. Freimer, Screening a large reference sample to identify very low frequency sequence variants: comparisons between two genes. Nature Genetics, 27, 435, 2001.
10. C. A. Jones, J. R. Sampson, and J. P. Cheadle, Low level mosaicism detectable by HPLC but not by direct sequencing, Human Mutation, 17, 233, 2001.
11. B. J. C. van den Bosch, R. F. M. de Coo, H. R. Scholte, J. G. Nijland, R. van den Bogaard, M. de Visser, C. E. M. de Die-Smulders and H. J. M. Smeets, Mutation analysis of the entire mitochondrial genome using high performance liquid chromatography. Nucleic Acids Res., 28, E89, 2000.
12. K. H. Hecker, P. D. Taylor, and D. T. Gjerde, Mutation detection by denaturing DNA chromatography using fluorescently labeled polymerase chain reaction products. Anal. Biochem. 272, 156, 1999.
13. L. Kruglyak, and D. A. Nickerson, Variation is the spice of life, Nature Genetics, 27, 234, 2001.
14. I. M. Buyse, P. Fang, K. T. Hoon, R. E. Amir, H. Y. Zoghbi. and B. B. Roa, Diagnostic testing for Rett syndrome by DHPLC and direct sequencing analysis of the MECP2 gene: Identification of several novel mutations and polymorphisms, Am. J. Hum. Genet., 67, 1428, 2000.
15. B. Hoogendoorn, M. J. Owen, P. J. Oefner, N. Williams, J. Austin, M. C. O'Donovan, Genotyping single nucleotide poymorphisms by primer extension and high performance liquid chromatography, Human Genetics, 104, 89, 1999.

Index

a
absorbance 30
accuracy 9, 40, 129
adjusted retention time 48
affinity chromatography 6, 59
agarose gel electrophoresis 45
alternative splicing 154
anion exchange 25, 76
autosampler injector 23 ff

b
back pressure 23, 44
background mutations 94
base pair cutter 161
bisulphite deamination 140
bisulphite treatment PCR 166
blind collection 132
BRCA1 100
BRCA2 100

c
calibration 110, 122, 126
calibration plot 111
cancer diagnostics 92
capacity factor 48, 50
cDNA library 149, 151
CEL I catalyzed cleavage 163
CFTR 99
check valves 19
chelating reagent 208
chromatograph 15 ff
– autosampler injection unit 15
– column 15
– data processing 15
– detector 15
– fragment collector 15
– mixer 15
– oven 15
– pump and gradient system 15
– solvent reservoirs 15
clamping 88
cleaning 121
cleaning procedures 26
cleavage products 148
cloning 119 ff, 162
cloning purification 8
column oven 27
column packings 45
column protection 26
competitive RT-PCR 152 ff
contamination 26, 120, 207
corroding stainless steel 207
CpG methylation 165
crosslinking 59

d
dead time 48
dead volume 17, 23, 41, 119
degassers 19
denaturation 83, 87
detection limit 95
detector noise 28
DHPLC 5, 8, 10, 27, 38, 81 ff, 214
DHPLC temperature 84 ff, 214
diagnostic 222
diethyl pyrocarbonate (DEPC) 133
differential display 155 ff
distribution coefficient 51
DNA denaturation 83, 87
DNA ladders 110 ff
DNA ligation 149
DNA methylation 164
DNA polymerase 91
domain temperature 83, 85, 87, 93, 96, 217
dwell volume 22 ff

e
eddy diffusion 55
EDTA 73, 211
electrospray ionization (ESI) 35
eluent 18 ff, 43
eluent back pressure 23, 44
enzymatic cleavage 161, 163
EXT1 103
EXT2 103
extra column effects 57

f
false negative 222
false positive 222
fidelity 218
fluorescence 32
– absorption spectrum 32
– detection 29, 39
– dyes 31
– emission spectrum 32
– intercalating 33
– tagging 33, 109
footprinting 175, 177 ff
four color detection 34
fragment collection 40 ff
frit 18, 25, 73, 75, 207
full loop injection 24
fully denaturing mode 71
functional groups 54

g
GC-clamp 88, 90, 96
gel electrophoresis 10, 45
gel permeation 58
gene expression 135
gene libraries 162
general detector 28
gradient 21 ff, 68 ff
gradient delay volume 23
gradient elution 68 ff
gradient formation 21

h
hairpin ribozyme complex 147
height equivalent of a theoretical plate (HETP) 48, 55
helical tendency 87
HEMA 63, 77
heteroduplex 81 ff, 89, 98
heteroduplex detection 10 ff, 81 ff
heteroplasmy 92
heterozygous 39, 217
HFE 114
high performance liquid chromatography 14 ff, 44
high pressure gradient 21
Holliday junction 178
homoduplex 81, 89
homozygous 39, 217
HPLC 14 ff, 44
HPRT 102
hydroxyapatite 7
hydroxyl radical footprinting 146

i
injection needle 212
injection valve 24
injector 22
insertion mutant 153 ff
INTERA2 155
intercalating dyes 181
internal standard 152, 154
ion exchange 6 f, 54, 58, 75
ion-pairing 6, 8, 67
ion pairing reagent 66
isocratic elution 21, 68 ff, 78, 123

k
k-ras 96

l
ligation reaction 162
linear flow rate 53, 55
liquid chromatography 43
longitudinal diffusion 56
loss-of-heterozygosity (LOH) 115
low pressure gradient 23

m
mass analyzer 37
– Ion Cyclotron Resonance (ICR) 37
– Quadrupole 37
– Sector (magnetic and electrostatic) 37
– Time of Flight (TOF) 37
mass concentration 39
mass sensitivity 93
mass spectrometry detection 34
mass spectrometry ionization 36
– Chemical Ionization (CI) 36
– Electron Impact (EI) 36
– Electrospray Ionization (ESI) 36
– Fast Atom Bombardment (FAB) 36
– Matrix Assisted Laser Desorption (MALDI) 36
Maxam-Gilbert sequencing 177
melt map 85 ff, 96
melting curve 215

metal contamination 72, 75
metal ion contamination 72, 75
methylated DNA 164
methylation specific PCR 167
mismatch endonuclease 163
mixer 22
MMAC1 98
mobile phase 42
molar absorptivity 30
molar concentration 39
molecular diagnostics 171
monolith polymeric columns 62
mRNA 135, 138, 141, 143
multi-domain 86
multipath 55
mutation detection 10 ff, 81 ff, 89, 98

n
NF1 99
nicked DNA 111 ff
nitric acid 208
non denaturing mode 71
nonporous 3
nonporous resins 59 ff, 77
nonporous silica 64
normal phase 58
nucleic acid enzymology 171
nucleic acid tagging 180

o
oligonucleotides 9, 109, 119
oligoribonucleotides 9, 135, 145

p
partial loop injection 24
partially denaturing mode 71
passivation of system 211
passivation treatment 207
PCR cycles 95
PCR enzyme 218
PCR fidelity 91 ff
PCR induced error 218
PCR infidelity 95
PCR noise 94
PCR products 8
peak capacity 53
peak pattern 38
peak resolution 48, 123
peak shape 38
peak width 48
phosphorylated and unphosphorylated DNA 173
piston seal 19
plasmids 110, 119, 162
plate number 53
plate theory 51
polyacrylamide gel electrophoresis 45
polymerase 95
polymerase fidelity 221
polymer-based anion exchangers 75
polymeric resins 59
polymorphism 81
polynucleotide kinase assays 173
pooling 219
porous resins 59, 77
porous silica 64
precision 9, 94, 129
preheat coil 27
primer extension 114, 169
primer optimization 88, 90
proofreading 91
proportioning valve 22
PTEN 98
pumps 19
purification 8, 118, 180

r
rate theory 53
recovery of material 122, 133
reservoir 22
resin capacity 54
resistance to mass transfer 56
resolution 93
RET 99
retention factor 48, 50, 54
retention time 48 ff
retention volume 49
reverse phase 6, 25, 54, 57 ff
reverse phase ion pairing 66 ff
RNA 8, 135
– conformation 144
– footprinting 146
– protein interactions 146
RNase 137 ff
rRNA 135, 138, 141
RT-PCR 109, 133, 136, 141, 143, 152

s
sample loop 24
scavenger columns 26
secondary structure 147
selective detector 28
sensitivity 9, 28, 33, 94, 221
separation column 16, 25
separation factor 48, 53
separation process 43
sequencing 119
silica-based anion exchangers 77

silica-based materials 64
single nucleotide polymorphism (SNP) 38 ff, 81 ff
slope detection 132
somatic mutations 92
splitting peaks 121 ff
stainless steel components 17, 121 ff, 207 ff
standard 9, 28, 38, 110
stationary phase 42, 54
strand-specific hybridization 180
STRs 119
substrate 59
system cleaning 207 ff

t

telomerase 139
telomerase assays 171
telomeric repeat amplification protocol (TRAP) 171
temperature domain 83, 85, 87, 93, 96, 217
temperature prediction 85, 88, 216
temperature titration 84 ff, 216
theoretical plate 53
theoretical plate number 48
thermal titration plot 84 ff, 216
thioredoxin-1 154
threshold detection 132
TILLING 103
timed collection 132
transcription 136, 141
triethylammonium acetate (TEAA) 67
tRNA 135 ff, 141, 143
TSC1 100
TSC2 100
tumor sample 98

u

uracil DNA glycosylase assays 174
UV detection 29, 39

v

van Deemter equation 56

y

Y-chromosome 99